Aleksandar D.
Rodić

Miomir K.
Vukobratović

DYNAMICS, INTEGRATED CONTROL AND STABILITY OF AUTOMATED ROAD VEHICLES

- research monograph-

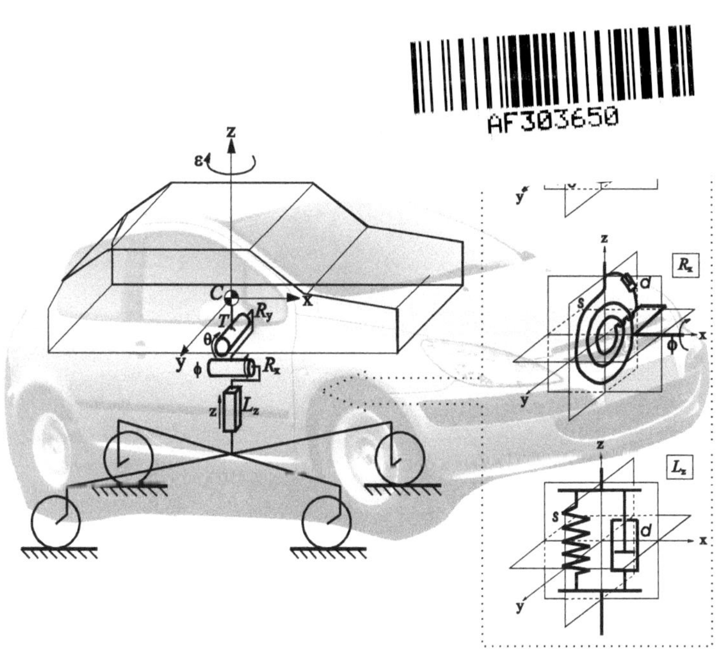

Aleksandar D. Rodić and Miomir K. Vukobratović

DYNAMICS, INTEGRATED CONTROL AND STABILITY OF AUTOMATED ROAD VEHICLES

ibidem-Verlag
Stuttgart

Die Deutsche Bibliothek - CIP-Einheitsaufnahme:

Ein Titeldatensatz für diese Publikation ist bei
Der Deutschen Bibliothek erhältlich

∞

Gedruckt auf alterungsbeständigem, säurefreien Papier
Printed on acid-free paper

ISBN: 3-89821-203-3
© *ibidem*-Verlag
Stuttgart 2002
Alle Rechte vorbehalten

Printed in Germany

PREFACE

The problems of driving safety, traffic efficiency and excessive contamination of the environment, brought about by a constant increase in the number of vehicles on the roads, have become very acute issues. These problems are especially pronounced in the industrially developed countries and urban areas. In order to overcome them, the automotive industry has to adapt itself to the traffic demands and improve the level of automation of traffic and vehicles, as well as their dynamic performance. The forthcoming evolution of road vehicles as transport means will go in the direction of improving their stability, active safety, maneuverability, ride quality, ride comfort, etc. Thus, the highest level of automation of road vehicles would be the synthesis of their fully active, integrated control systems.

From the point of view of the system control design, motion control and control of the vehicles dynamic performance represent a very delicate and challenging task. Namely, it is necessary to ensure the control simultaneously satisfies several requirements such as global stability, ride quality, ride comfort, minimal dynamic loads of the mechanical subsystems, low energy consumption, etc. The choice of the appropriate control strategy and the way of its realization represent a crucial problem, the solution of which demands sufficiently deep knowledge of the dynamic behavior of the road vehicle in various motion conditions. In the course of the last two decades, much attention has been paid to the problems of vehicle control and traffic automation. The main stimuli to this activity were of technical, social, ecological and, normally, economical nature. Hence, intensive research in this field represents an imperative of the technological progress in this industrial branch, as well as a challenge to the teams of researchers and designers, working in the field all over the world.

The objectives of writing this book

There are several good books dealing with modeling and automotive control. However, the control algorithms described in them are prevailingly based on the so-called decentralized principle using most often simplified, planar vehicle models and the common control techniques - optimal or robust. Thus, up to now, there has been no a text considering the centralized approach to practical vehicle stability analysis. Because of that, the goal of this book was to stress out the importance of the knowledge of vehicles dynamics and of the benefits of implementation of the integrated control in the advanced vehicle controllers, as well as the importance of the systems stability analysis in the synthesis of dynamic control laws. The basis of the research presented in the book is a complex, spatial dynamic model, which describes the most pronounced dynamic effects in the road vehicle systems. Based on the vehicle integral dynamics, two novelties were introduced:

- integrated dynamic control of road vehicles based on centralized control approach, and

- practical stability analysis of the vehicle system.

Organization of the book

The monograph consists of several topics that have been organized in five chapters:

- Introduction

- Modeling of Vehicle Dynamics

- Control of Automated Road Vehicles

- Stability Conditions of Road Vehicles

- Summary and Conclusions

Each chapter begins with a short abstract informing the reader about the matter to follow.

Chapter 1 describes the objectives and importance of research in the field, along with the definitions of some basic notions to be used later in the text.

Chapter 2, dealing with modeling, contains a rather comprehensive survey of the results in the given research area, so that this chapter can be used as a textbook on road vehicle modeling.

Chapter 3 describes a novel concept of the so-called integrated dynamic control of robotized road vehicles, presented for the first time in this book and in the recently published journal papers.

Chapter 4 gives a novel contribution to the analysis of the vehicle practical stability on the basis of relatively complex spatial dynamic vehicle model, as a strongly coupled mechanical system, especially in the course of high-speed maneuvers. The most important theoretical results are accompanied by the corresponding simulation experiments, needed to achieve clarity of the presented subject matter and emphasize the aspects of its practical implementation.

Chapter 5 gives a brief conclusion concerning elaborated matter.

Background and potential readership

The material presented in this research monograph is founded on the application of the software developed in the MATLAB programming background. The software for simulation, control synthesis and stability analysis makes the basis for development of the potential real-time control software. In the writing about the dynamics of active spatial mechanisms, dynamic control of large-scale systems and their practical stability tests, we drew upon the material represented in the first monograph series *Scientific Fundamentals of Robotics*, written by Professor M. Vukobratović and his associates and published by Springer-Verlag in the period 1982-1986. Besides, the source of knowledge was the results from numerous articles published in the open literature dealing with automotive control engineering. In addition to its basically research character, this monograph, thanks to the presented results in the domain of integrated dynamic control intending to be applied in road vehicle integrated dynamic control, has also certain features of an advanced textbook. Hence, it is primarily dedicated to post-graduate students, and particularly to those interested in the dynamic control based on the overall dynamics model of the road vehicle. Besides, the book can be used by the lecturers at technical faculties, teaching the engineers to whom the road vehicle dynamics is not an ultimate goal but a means towards the solution of the problems in road vehicle control. Of course, the monograph may serve as a useful handbook to the researchers working in the field.

We think that this research monograph will enable the reader to gain a sufficient knowledge for his further work on the problems of dynamics and dynamic analysis of road vehicles and for implementation of theoretical approaches into practice. We also believe it will help the readers in their future work on the problems of dynamic control of road vehicles.

How well we have succeeded, it remains to be judged on the basis of the use of this book as an advanced book in teaching practice, as well as in the research and development units for applied dynamics and dynamic control of road vehicle systems.

The authors express their gratitude to Professor Dr. Luka Bjelica for rendering the book into English. Besides, the authors specially aknowledge to the Alexander von Humboldt Foundation (Germany) that granted the research project based on the scientific results presented in this book. Also, the authors express their sincere gratitude to Professor Dr. Eckehard Schnieder and his staff from the Institut fuer Verkehrssicherheit und Automatisierungstechnik (TU Braunschweig, Germany) that enable and support the realization of the project.

March 2002,   The Authors
Belgrade

Contents

List of Figures

Chapter 1

Introduction

Abstract

In this section readers get briefly acquainted with academic problems concerned in the monograph. First the subject of research is defined, then the state of art in control of road vehicles is briefly described, and the still existing problems in this field of research will be stressed out, too. Afterwards, research objectives and applied methodology are presentd. At the end of this section the topics of the monograph as well as the significance of the active control systems operating with road vehicles will be briefly explained.

1.1 Description of the Academic Problem

The problems of driving safety, traffic efficiency and excessive environmental contamination, brought about by the constantly increasing number of vehicles on the roads, represent very serious issues of our time. These problems are especially acute in the developed countries and urban areas. In order to overcome such a grave state and avoid even worse prospects, the automotive industry has to adapt itself to the traffic demands and enhance the level of traffic and vehicle automation, as well as of the vehicle dynamic performance. The directions of the forthcoming evolution of road vehicles as transport means involve improvements of their stability, active safety, maneuvering ability, ride quality, ride comfort, etc. Thus, the highest level of automation of road vehicles is the synthesis of the appropriate, fully active, integrated control systems. Apart from attaining higher controllability, safety, and ride quality, the objective of vehicle automation is to introduce maximally automated control systems, to replace human factor as much as possible, as it represents an element of limited control capabilities. Full automated control of road vehicle motion and its dynamic performance is mainly concerned with the problems of Intelligent Transportation Systems (ITS). Some remarkable practical results have been achieved in control of vehicle dynamics in particular directions such as longitudinal, lateral, and vertical, as well as in the roll, pitch, and yaw directions. Significant progress in this field has been made thanks to the advances in the microprocessor and sensor technology and technological progress in the field of the development of driving units. Thus we can speak about a new generation of vehicles which are essentially different from the conventional ones involving only passive control elements. They possess active control systems such as active suspension system, active steering system, active traction system, active braking system, systems with variable geometry, hybrid power drivetrain, etc. These systems enable microprocessor control of motion and performance of road vehicles, improving significantly their dynamic characteristics and stability. From the point of view of control system design, control of motion and system's dynamic performance represents a very challenging and delicate task. In that sense, control should execute simultaneously a multi-criterion task of ensuring vehicle's stability, ride quality, ride comfort, minimal dynamic loads of the mechanical subsystems, low energy consumption, etc. The choice of the best control strategy and the way of its realization represents a crucial problem, solution of which demands a sufficiently good knowledge of the dynamic behavior of the vehicle under various conditions of motion. In the course of the last two decades much attention has been paid to the problems of vehicle control and traffic automation. The reasons for this are of technical, social, environmental, and of course, economical nature. Hence, intensive research in the field is an imperative of the technological progress in this industrial branch, as well as a challenge to the teams of engineers and designers engaged in it all over the world.

1.2   Subject of the Research

The aim of this research monograph is to investigate the problems of active integrated control of road vehicles as the complex systems interacting with their dynamic environment. Based on the analysis of the results previously achieved in the domain of automotive engineering, the monograph presents a new concept of vehicle control which may be regarded as an alternative approach to the contemporary research in the field. During the last two decades, some remarkable results have been achieved in solving the problems of vehicle's dynamic control in particular directions of motion (longitudinal, lateral, yaw, roll, and perpendicular to the road surface). The current research is concerned with the unresolved problems in automated control of vehicle performance. In this sense, it is necessary to point out certain drawbacks and imperfections of the existing solutions and clearly indicate the unsolved problems in this field of engineering. The main contribution of this research monograph represents the proposed, essentially new, algorithmic solutions of road vehicle automatic control using active control systems as an integrated system of guidance. The new approach to control of the integrated vehicle dynamics is founded on the centralized principle yielding the design of a vehicle autopilot (full automated control). Hence, the research subject of this monograph could be defined as the development of new control schemes and dynamic control algorithms based on the novel strategy of road vehicle integrated control using various types of active systems. Investigations concerning the synthesis of road vehicle active control can be classified in two basic groups:

(i) Investigations in the field of designing fully automated control (autopilot synthesis) of road vehicles as complex dynamic systems without driver's manual participation.

(ii) Investigations in the field of driver-assisted control of interactive vehicle dynamics.

In this monograph we are going to deal with the problems of the first group. Thus, we shall consider the problems of stability and control synthesis concerning the road vehicles operating with automated highway systems, assuming the existence of a fully automated traffic regime. A complex vehicle autopilot controller, representing the hybrid neuro-dynamic controller, will be described in Chapter 3. Investigations in this field of control engineering represent a very attractive field for researchers and can be commercially interesting for the automotive industry. In the forthcoming time, thanks to the infrastructure of highway systems, there will exist highly automated traffic regimes involving highly automated road vehicles. For these reasons, design of advanced control systems for road vehicles is a matter of technological prestige in the automotive industry for the reasons of high criteria of driving safety, ride quality, and ride comfort that have to be satisfied in this industrial branch. Before summarizing the contemporary achievements in the field of automated control of road vehicles it would be useful to define the task of control. It can be formulated in the form of an essential question: "Can we design such control that will ensure simultaneous stability of the dynamic behavior of a dynamic system (transport means) in the three basic motion coordinates longitudinal, lateral and yaw?" At the same time, care has to be taken to ensure a high quality of the vehicle behavior with respect to the road surface (roll, pitch and heave motion). This is an extremely complex and delicate task because of the existence of various internal and external perturbations of stochastic nature acting on the system. The following perturbations should be primarily taken into account: variation of the friction coefficients of tire pneumatics, changes in the road surface profile, appearance of aerodynamic forces of resistance due to air streaming around the vehicle body, variation of the kinematic and dynamic parameters of the system, time delays in the vehicle actuators, etc. All these factors can destabilize the system and, because of that, a designer who is to design a controller for a stable vehicle behavior must take into consideration all of them.

1.3 State of the Art in Control of Vehicle Dynamics

In the last several decades, significant advances have been made in the theoretical and experimental research in the field of our present concern. The problems of control of longitudinal and lateral interactive dynamics and active safety have been approached in the following three ways:

(i) By the synthesis of control laws based on the implementation of active, independent four-wheel steering (I4WS),

(ii) By applying direct yaw moment control via active braking system or traction control system (TCS) and,

(iii) By using the integrated control combining the I4WS and independent four-wheel driving (I4WD), implementing direct yaw moment control (DYC) algorithms.

Some of the most significant results that have been achieved in this field will be presented in the text to follow.

In the recent years, several car manufacturers have developed car prototypes equipped with non-conventional steering systems based on a 4WS configuration [1] - [3]. The additional degree of freedom (DOF) offered by rear wheels steering can be exploited to improve the vehicle dynamics by means of suitable automatic controllers, while leaving the primary function of path tracking to the human driver. Such improvements include better maneuverability at low speed, enhanced stability at high speed, reduction in sideslip angle and elimination of lateral force disturbance. In view of the control scheme applied, the solutions proposed for 4WS cars can be grouped in three main categories: (i) feedforward, (ii) feedback rear-active, and (iii) feedback full active. In feedforward controllers, rear wheels steering is subordinate to the action on the front wheels, which are directly commanded by the driver. On the contrary, feedback configurations make use of the measured vehicle variables. In the rear-active schemes, the driver maintains mechanical control of front wheels as in conventional cars, and the controller acts only on the rear axle. Full-active controllers yield a coordinated action on the four wheels in response to the driver's commands or external disturbances. Examples of feedback control systems can be found in [2], [4]. Recently, a feedback full-active controller for a 4WS vehicle has been developed using the μ-synthesis techniques [5], [6]. The following main objectives have been pursued:

(i) achieving good lateral displacement tracking performance in response to the driver's command;

(ii) reducing the effect of external disturbances on the lateral dynamics; in particular, the vehicle should be able to autonomously respond to side wind perturbations;

(iii) minimizing the sideslip angle in all maneuvers, including high-speed turns and fast lane change;

(iv) preserving stability and performance robustness in the face of uncertainties in the tire cornering stiffness coefficients.

Design of the most of existing control schemes relies upon the simplified linear models of the vehicle lateral dynamics. These approximative models are suitable for the controller design, but they neglects some important aspects of the interaction between tires and road surface. The phenomena like roll and pitch dynamics and a sophisticated description of the nonlinear tire characteristics have to be included into consideration in the control synthesis, too.

Some analyses have been carried out concerning the vehicle dynamics in critical cornering, considering the vehicle stability from a viewpoint of the quasi-static yaw moment balance [7]. However, it was difficult to express sufficiently well the vehicle dynamic stability by these approaches. The state-plane method is generally considered effective as a method to analyze stability of a dynamically nonlinear system. Several attempts have been made using this method, focusing on the vehicle lateral dynamics, i.e. on the sideslip angle and yaw velocity. A new approach to the vehicle stability in critical cornering, using the "phase-plane method" was proposed by Inagaki et al. [8]. They have taken into consideration the vehicle sideslip angle and its angular velocity as the state variables. Then, it has been verified that the direct yaw moment control system, with active braking using the algorithm based on the phase-plane analysis, makes the vehicle stability much higher in critical cornering by an experimental car.

The efforts directed toward diminishing the occurrences of traffic accidents and improving the capability of avoiding them have resulted in a number of structure control systems that are presently available in the market. It has been expected that active steering systems such as rear wheel steering, adaptive power steering and vehicle stability control systems, etc., would contribute to solving active safety problems. Therefore, it was necessary to establish the appropriate assessment of handling quality or active safety in objective and quantitative ways, so that it would be possible to optimize many design parameters and control strategies in an early stage of vehicle development. In that sense, a closed-loop system "driver-vehicle model" for handling and stability quality was analyzed in [9], in order to evaluate appropriately the effectiveness of rear-wheel steering systems. A linear preview control model of driver's steering maneuver was assumed in the cases of lane change and straight line running against the crosswind gust. Then the evaluation functions of handling quality were defined to be weighted mean square variables as performance indices. Furthermore, in order to specify optimum vehicle response parameters, the relationships between the evaluation function, i.e. the driver-vehicle performance and vehicle design parameters were analyzed, comparing the typical systems of rear-wheel steering of feedforward and feedback types.

The 4WS systems with yaw rate feedback control have been investigated most often in order to improve the handling and stability of the vehicle and reduce influence of disturbances on its lateral motion. However, the vehicle lateral dynamics changes dramatically with the change in friction coefficient of the system vehicle-road surface, which can be implemented in real time with a few on-board sensors in a 4WS system. The fact that steady-state yaw rate gain on a slippery road is smaller than that on high-friction roads was utilized by some authors (e.g. [4]) to estimate the friction coefficient from the ratio of the actual yaw rate to the linear model yaw rate on the high-friction road. As second, a feedforward compensator was basically determined so that the steady-state and transient vehicle slip angle remained zero while the road and driving conditions were changed. As a result, a compensator was formulated as a first-order transfer function whose coefficients were dependent on the estimated friction coefficient and lateral acceleration. Next, the yaw rate feedback system was designed so that the overall system achieves good performance under various road conditions. For this purpose, Wakamatsu and co-workers from the Honda

Motor Co. [4] adopted an Internal Model Control (IMC) structure incorporating a linear compensator with a nonlinear vehicle model. The reason for adopting IMC as a feedback control was its easy implementation for compensating the nonlinearities and for adopting the parameters in the vehicle model. The linear compensator was designed by the μ-synthesis, one of robust control techniques, in order to ensure the robust stability and performance even if unexpected modeling and estimation errors existed in the feedback system.

A new approach to design of the active front-wheel steering system, based on the "Robust Model Matching" (RMM) algorithm, which enforces desired frequency characteristics between the driver steering angle and vehicle yaw rate, was proposed by Tagawa et al. [10]. This resulted in a vehicle having the same frequency characteristics in the face of parameter errors and nonlinear dynamics. But this control system can become unstable when the vehicle speed increases and the lateral force exceeds its limit. In order to prevent occurrence of such a situation the controlled quantity is modified by introducing a new function that consists of the vehicle speed, steering wheel angle, and the maximum lateral force of the vehicle.

Hirano and Fukatani [5] have proposed a new logic for a robust active rear steering (ARS) system. It maintains good control performance even if the vehicle parameters and/or road surface conditions are changed. This system made the vehicle to follow the desired vehicle model by the state feedback of both yaw rate and sideslip angle. To utilize noise yaw rate sensor and also to keep high robust control performance a model matching controller (MMC) with frequency-dependent gain characteristics was designed by μ-synthesis. To estimate the sideslip angle using a noise sensor, a new observer applying frequency filter was introduced. The control law for this system was derived so that the vehicle sideslip angle becomes zero, both in steady and transient states. Feedforward ARS can improve vehicle handling during cornering and lane change, but it cannot stabilize the vehicle in the presence of external disturbances. To cope with this problem, a yaw feedback ARS system was developed at the end of the 1980s. However, because this logic is solely derived from an inverse second-order model of vehicle dynamics, robustness of the control performance in the presence of changes in vehicle mass, tire characteristics and/or road surface conditions cannot be guaranteed. Moreover, because this system needs precise yaw rate signal in all the frequency range, rather expensive yaw rate sensor is required. There is another shortcoming of the above systems. To compensate for the decrease in the steady-state yaw rate gain by the ARS control the front-steering gear ratio should be changed to a higher value. This results in an unnecessary increase of the gain of yaw-rate response in the range of high frequencies, and causes an immoderate yaw movement when fast steering input is given. To avoid this problem in the actual yaw rate feedback ARS system, the yaw-rate feedback gain should be restricted to a value that is considerably smaller than the theoretical value, especially at high speeds. Therefore, an unsatisfactory control performance can be obtained with the actual systems. On the other hand, an ARS control law based on the model following control by the state feedback using LQR controller, has been proposed. This system feeds back yaw rate and sideslip angle to let this actual value follow a desired model. Also, an observer was utilized to estimate sideslip angle from the yaw rate. However, there also remained the problem of robustness as well as the necessity of having also a high-precision sensor in these systems. Furthermore, it is required that the feedback gain of the controller and the observer's parameters vary according to the vehicle speed in actual applications. Results of various simulations and actual tests have proved the great effectiveness of the new ARS system.

The steering system can be considered as the most important driver-vehicle interface that provides the driver information on the vehicle motion being controlled. This implies that the vehicle handling qualities can be improved by actively transmitting to the driver information on the vehicle motion and/or external disturbances as artificial supplementary steering torque in the steering system, i.e. artificial kinesthetic information. With this in view, Yuhara and Horiuchi [11] have conducted their recent studies. Their research efforts resulted in the realization of a driver-assistant steering system that can improve the task performance both in the straight-line keeping and lane-following on a curved road (a tracking task). The system aimed at achieving good handling qualities by means of active kinesthetic information offered by the steering system, i.e. good task performance and low driver's mental workload. The driver always remains in the loop and is always able to provide primary steering commands. This system contains a virtual model that represents the yaw rate response of the vehicle. The virtual vehicle model has its response set at higher speed than that of the actual vehicle, so it should work as a reference model of the actual vehicle. The advanced steering system feeds back artificial supplementary steering torque as well as self-aligning torque to the steering wheel. The artificial supplementary steering torque is generated so that it is proportional to the difference between the yaw rate of the actual vehicle and that of the virtual vehicle model. Therefore, through the steering wheel, the driver receives a supplementary steering torque needed to make the actual vehicle follow the virtual vehicle model as kinesthetic information. The proposed steering system may be considered as another form of model-following control.

Ackermann, Sienel et al.,[12] proposed a robust control law based on feedback of the yaw rate into active front steering. Practical driving tests showed essential safety advantages in respect of the effect of side wind and braking with asymmetric road adhesion. On the other hand, some authors worked on the feasibility of integrating 4W-steering and active 4W-driving control systems using a well-known method [3]. However, such control can guarantee only robust control stability, but cannot ensure robust performance.

Recently, many additional systems for vehicle handling and stability, not only at constant speed turning but also at braking and traction, have been proposed. Remarkable achievements with these systems generally depend on the tire force characteristics, especially at high lateral and longitudinal accelerations. Tire force conditions strongly affect the control system performance. There are two known control methods for the tire force coefficient, one is direct yaw moment control method (DYC) and the other is the robust active four wheel steering (4WS).

Four-wheel steering systems have been subject of research and development in the last two decades with the aim of improving handling and stability characteristics. Especially diverse active control systems have been developed for rear-wheel steering by feedforward and/or feedback compensation for sideslip motion in the vehicle body. Although the improvement of handling and stability results in quite a good effect on driver's steering maneuvers (the so-called closed-loop performance of the driver-vehicle system), it is limited to a linear dynamic system because of the decrease of cornering stiffness of tires during a high lateral acceleration. The systems to control yaw moment using driving/braking forces have been researched and yielded also improvements of the vehicle handling and stability. One of them is active traction control system of each wheel through the feedback of state variables, such as yaw rate. The other is active braking control system through the feedback of state variables, such as yaw rate and/or sideslip angle in the vehicle body. These active control systems can guarantee that the yaw moment directly compensates for the vehicle yaw motion, not only in the linear ranges but also in nonlinear ranges of tire performance. Therefore, the yaw-moment control systems are expected to suppress the deterioration of the steering control effects in nonlinear or large lateral acceleration ranges.

Nagai and Yamaka [13] proposed an integrated control system for active rear-wheel steering and direct yaw-moment control. The designed control system is an MMC which makes the vehicle follow the desired dynamic model by using the state feedback of both yaw rate and sideslip angle. Although this integrated MMC was designed on the basis of linear control theory, it can greatly improve the handling and stability and show the expected robustness at the changes of the vehicle parameters such as the cornering stiffness and vehicle's mass, even in large ranges of lateral acceleration and in low tire-road friction ranges.

Aiming at overcoming the drawback of 4WS due to tire saturation property of lateral force to sideslip angle, a number of studies used some sophisticated nonlinear control theories to introduce the 4WS control law. However, as sufficiently deep insight into the physical characteristics of tire and vehicle dynamics is not necessarily taken into account in the control theories used, the disadvantage of 4WS, especially in improving near-limit handling performance, has not completely been cleared away. On the other hand, there have been some encouraging recent studies on direct sensing of tire friction coefficient and lateral force, which might be essential information for improving the near-limit performance of vehicle handling. Furukawa and Abe [14] introduced a new control law for direct yaw moment control. This effective method produces a control yaw moment depending upon tire longitudinal forces, as a cooperative control with 4WS to compensate for its drawbacks during saturation of tire lateral force. The control moment in yaw direction to be produced by tire longitudinal forces is determined so that the vehicle motion follows the predetermined model response even in the range of saturation of tire lateral forces, to prevent it from falling into catastrophic motion. For this control synthesis, the estimation of tire lateral forces becomes essential. The vehicle with such chassis control has simple on-board-tire-model by which the tire lateral forces during the vehicle motion are estimated and used for control. The model response is set so that the sideslip angle of the vehicle converges to zero immediately, even in nonlinear maneuvering conditions. The vehicle is assumed to be equipped with a linear feedforward zero-sideslip 4WS.

In addition to pure longitudinal slip control systems, newest technologies consider both longitudinal and lateral slip angles. These Dynamic Driving Control (DDC) systems monitor also the lateral dynamic driving behavior and, in contrast to the longitudinal slip control systems, brake and engine management intervention may be effectuated without the driver's activity. The ABS and TCS can recognize critical longitudinal driving situation and control the longitudinal slip ratio. If at least one of the wheels starts to lock or spin, the controller will adjust the longitudinal slip ratio to the maximum possible longitudinal tire force (about 10 % of the longitudinal slip ratio). However, maximum lateral vehicle stability will not be reached at this operating point, wherefore ABS and TCS cannot be used to optimize the vehicle's lateral driving behavior [15]. The DDC supports the driver in critical driving situations. If the vehicle enters a critical situation, the control algorithm will stabilize the vehicle by certain individual brake forces. In case of understeering behavior, main brake intervention takes place on the inner rear wheel. The rear brake force causes a primary yaw torque and reduction of the rear lateral tire force. Both effects help the vehicle's cornering. In case of oversteering behavior, main brake force on the outer front wheel helps to stabilize the vehicle. The intervention produces a primary yaw torque and reduces lateral tire force at the front side. These effects prevent critical oversteering driving situations. The DDC system designed by Pruckner and Seamann [16] uses both yaw velocity and sideslip angle to control the lateral vehicle behavior. The use of only yaw velocity can cause wrong identification of the vehicle state and the controller alone cannot sufficiently overcome some critical driving situations. Pruckner and Seamann proposed a state controller algorithm that uses a flat double-track vehicle model with integrated HSRI tire model [16]. This model is linearized about the actual working point during each calculation step. The control algorithm is divided into three subsystems: (i) state identification, (ii) state observation, and (iii) state control. The identification serves the driver's wanted state, observation serves the actual vehicle state and the control system compares these two states. Depending on identified understeering or oversteering driving situation, brake force is applied to the inner front wheel or the outer rear wheel. The state controller was designed to consist of two subsystems, Ricatti controller and slip controller. Together, they make the DDC vehicle's controller intended to improve the control system stability.

1.4 Still Existing Problems

On summarizing the published research contributions in the field of automotive engineering we can identify the following problems and unresolved issues:

- In spite of the fact that significant improvements of active safety and vehicle controllability have been achieved, there is still lack of a satisfactory solution to the problem of vehicle's simultaneous stability in the six main[1] directions of motion. The reason for that is in the strongly coupled vehicle dynamics and in extremely nonlinear nature of tire pneumatics, the issues being rather difficult to control by the known methods from linear control theory. Strong coupling between the longitudinal, lateral and yaw motions, as well as between heave, roll and pitch motions, especially during cornering and driving at high speeds, represent a serious problem for control synthesis, which has not been solved yet in a satisfactory way.

- Controllers of interactive vehicle dynamics are usually synthesized on the basis of planar or linearized simplified models. These models describe the vehicle dynamic behavior in particular directions of motion. The commonly used so-called "single-track" planar model is given as reference for the control of lateral and yaw vehicle dynamics. As is well known, linear models are good approximations of real systems only in a limited range of values of the input and state variables. Outside this range, the above models are only rough approximations. This fact has certain consequences in control applications. However, the advantages provided by linear models because of their simplicity in application of the well-known techniques and methods from linear control system theory are really obvious. For this reason, the pragmatism in modeling that has been shown by numerous researchers that have been using these models for implementation of appropriate control techniques (linear optimal regulator in state space or in frequency domain, H^∞-control, sliding mode control, μ-synthesis, model matching control, etc.) is understandable, but one should be careful with their implementation.

- When approximate system models are used, many dynamic effects such as elastic modes of vehicle structure (mechanism), Coulomb's friction in the rotational and sliding elements of the mechanism, time-delay in actuators, stiction in the hydro-cylinders, etc., remain unmodeled. In that way, in order to overcome the structural uncertainties of the modes in control synthesis, these effects should be additionally compensated by using some of robust control techniques. In view of the above unresolved academic problems in this research field, the proposed novel approach is an attempt to overcome some of them using different control principles and knowledge of applied mechanics and theory of stability.

1.5 Research Objectives and Applied Methodology

The research study concerning integrated control of vehicle dynamics shall be elaborated in the monograph. A new strategy for the integrated vehicle con-

---

[1]These are longituadinal, lateral, and yaw directions, as well as heave, roll and pitch directions of the vehicle body.

trol will be proposed for the first time. In that sense, we give a novel design of an advanced hybrid neuro-dynamic vehicle controller in longitudinal, lateral and yaw directions, as well as in the roll, pitch and vertical directions. The proposed control strategy is based on the principles of centralized controller synthesis, which has been established in the previous authors' publications [17] and [18]. Various types of active control systems of road vehicles, as well as power drivetrains will be considered for this purpose. A spatial, nonlinear, dynamic model of road vehicle will be assumed as a basis for the synthesis of control algorithms. Besides, several active systems such as I4WS, ABS, TCS, active suspension system (ASS), etc., will be used to attain the possible higher degree of control decentralization and better maneuvering characteristics. The new approach to active control of the vehicle interactive dynamics differs from the other known approaches in respect of the requirements concerning implementation of a more accurate nonlinear dynamic model of the system. In that way, the most of relevant effects that are excited in the system at high speeds will be taken into consideration, as well as its dynamic behavior in the case when the input and state variables change in a very wide range. The necessary robustness of control laws with respect to structural and parametric uncertainties will be achieved as one of most important goals in the monograph. This goal will be attained by using the combination of model-based and knowledge-based control algorithms. The basis for the synthesis of the new hybrid, model-based - knowledge-based vehicle controller, makes a dynamic controller with the primary role to compensate for the basic dynamic effects due to the changes of the driving conditions. Dynamic effects within the system are the consequences of increasing longitudinal and lateral accelerations as well as of the aerodynamic resistance forces acting upon the vehicle body. Intelligent control techniques will be applied as a complement to the dynamic controller, aiming at additional compensation of unmodeled effects due to model uncertainties. The adaptation to the variable motion conditions will be achieved by scheduling of control gains in the scope of the designed dynamic controller. Bearing in mind what we said previously, the main goal of the investigation dealt with in this monograph would be the synthesis of a hybrid neuro-dynamic controller. Its role is to ensure the necessary system stability and desired dynamic behavior of the road vehicle simultaneously in the main six directions of motion. It is expected that the future neuro-dynamic controllers will be designed so to ensure high stability performance of the road vehicle as well as to improve the quality of dynamic behavior in stationary and transient regimes of motion in comparison with other applied control systems.

In our proposal we apply the principle of centralized dynamic approach to control synthesis. This means that the entire integrated vehicle model should be applied for the synthesis of dynamic control algorithms. With this in mind, distributed hierarchy control strategy [19] is applied at two control levels: tactical and executive. In designing the hybrid vehicle controller, as well as in the synthesis of the state observer we use the spatial complex vehicle model. The synthesis of robust control methods based on the simplified linear models has a consequence that the stability of a dynamic system can be guaranteed only in particular directions and only under specific conditions. In the investigations presented in this monograph, the robustness of the control system will be attained in such a way that control gains will be calculated using automatic, computer-oriented procedure for testing practical stability conditions of the considered closed-loop system. It will be explained in the next sections of this monograph that the goal of checking practical stability conditions is directed towards the determination of the best parameters for both control and design.

Also, the problems of the synthesis of control for automated road vehicle will be considered as the so-called "contact tasks", in the way it has been done with similar dynamic systems [20]. Namely, in this approach, the overall behavior of the road vehicle dynamic system is analyzed together with its dynamic environment. For this purpose in Chapter 2 we present a specific spatial model of road vehicle which comprises both the rigid-body dynamics of the road vehicle and dynamic characteristics of its "environment". This model will be used in the synthesis of dynamic control laws that will be dealt with in Chapter 3. Let us return now to the notion of the so-called "contact task". Under this term we understand the solving of the problem of control of any dynamic system (such as for example ground vehicle, manipulation robot, aircraft, etc.) which in the course of motion interacts with its dynamic environment. At that, the environment can be of very diverse nature, i.e. it may have different characteristics of behavior in dependence of its physico-chemical properties. If we consider an airplane as a characteristic dynamic system, its dynamic environment is the air space that resists motion through it and at the same time generates the reaction forces on the airplane body. Under typical "contact task" in case of airplane we can consider the airplane landing, when it establishes contact with the runway surface. In the case of a ship, dynamic environment is the water the ship is immersed in and through which the ship body moves. The water possesses the corresponding viscous properties and as such hinders the motion by exhibiting resistance, thus decelerating sailing. Further, with railway trains, the environment represent the steel railway tracks, which have also their elastodynamic characteristics. Bearing all these examples in mind, under the notion of "contact" we mainly understand the direct touch of the object with its solid support. As we have already mentioned, the airplane in its landing comes in contact with the runway surface. A manipulation robot engaged in a machining operation by its end-effectors (terminal link of the mechanism's chain) performs a contact task. In this case, the tool acts against the corresponding resistance force of cutting, sliding resistance, ground reaction, etc. Finally, returning to road vehicle as the object of control considered in this monograph, it can be stated that the vehicle in its motion on the road surface interacts with the ground through its tire pneumatics. In the light of the above consideration, the road vehicle is looked at as a characteristic large-scale dynamic system performing the motion on the basis of the interaction with its dynamic environment. In its motion, the vehicle has to overpower the reaction forces (resistance) which tend to stop the vehicle or change its direction. The inertial and elasto-damping properties of tire pneumatics, viscous shock absorbers and spring dampers have a predominant influence on the intensity of the so-called contact load forces. These forces act in the direction perpendicular to the road surface plane. The intensities of these forces determine also the values of the coefficients of tire rolling and sliding friction and the amplitudes of their longitudinal and lateral forces. In a word, the motion of the vehicle and its stability depend on the way the contact of the tires with the surface on which the vehicle moves, is established. In the following considerations (while not departing much from the reality) we assume that in the sense of mechanics the vehicle is a rigid body (or a system of rigid bodies) which relies on the support through the so-called suspension system. Having the above in mind, we can conditionally assume that the vehicle suspension system and the ground on which the vehicle moves make in a wider sense the vehicle's "dynamic environment". Such functional decoupling of the road vehicle into vehicle body and its "environment" will be effectively used in our further discussion, in deriving the appropriate model, and in the synthesis of dynamic control laws. The above approximations can be justified by the following facts. The road on which the vehicle moves has certain physical properties. Assuming that the vehicle moves on a solid and not on a pliant (sand, gravel, earth) ground, the dynamic characteristics of the suspension system have a dominant influence on the dynamic behavior of the vehicle with respect to the support. The road surface itself can be practically considered as a kinematic constraint to the vehicle motion.

In view of the above, the main objectives of this research monograph can be summarized as follows:

(i) Development of the road vehicle dynamic model based on a spatial, non-linear dynamics of vehicle structure;

(ii) Synthesis of an advanced integrated control system intended for active control of dynamic performance and motion of the road vehicle using various types of active systems;

(iii) Determination of scheduling control parameters based on the implementation of an automatic procedure developed for checking practical stability conditions, in order to achieve the necessary robustness of the applied control schemes to the parametric uncertainties;

(iv) Application of intelligent control techniques in the frame of the designed controller, aiming at increasing the robustness of the proposed feedback control schemes to the inaccuracies of the applied model.

1.6 Active Systems in Road Vehicle Control

Good controllability of a road vehicle in every instant of its motion and high maneuvering capabilities are crucial requirements that must be satisfied by automated road vehicles in order to justify their existence and functioning. High steering performance of the vehicle can be ensured only by applying the so-called active control systems functioning in the frame of vehicle structure. In contrast to automated road vehicles, conventional vehicles have almost exclusively passive control systems. The design solutions for active and passive systems of road vehicles are not of our present concern. This has been the subject of the abundant writings and numerous commercial solutions that have already found their application in the automotive industry. In Chapter 2 we present several characteristic examples of active systems suitable for the application with automated road vehicles. These systems will be described by their functional schemes. This will meet our purpose of designing the corresponding models needed for the synthesis of the appropriate dynamic control laws.

Potential applications of active systems in the control of dynamic systems yield the capability of active systems to adapt themselves to the different levels of external forces, inaccuracies, and perturbations. However, the implementation complexity and maintenance costs can be considered as shortcomings of active systems.

A basic characteristic of all active dynamic systems is that they can change their dynamic behavior with time under the influence of external energy. In this way it is possible to change the dynamic parameters (moments of inertia, damping and stiffness coefficients), geometrical characteristics and kinematic quantities (velocity and acceleration) of the mobile elements. These possibilities can be in different ways used with road vehicles to achieve a better controllability of motion and system performance. The application of various types of active systems yields full or partial automation (robotization) of the vehicle. Thus the vehicle acquires characteristics of a robot, i.e. of the object that can move autonomously along a predetermined trajectory without the involvement of an operator.

Conventional road vehicles represent passive dynamic systems. They comprise various functional subsystems such as drive (traction) system, braking system, suspension system, steering system, electrical system, fuel system, communication system, senzor system, etc. Of special interest are here the following four systems: traction system, braking system, steering system and suspension system. Characteristics of these systems have a direct influence on the vehicle maneuvering abilities, motion stability, ride safety and ride comfort. Technical capabilities of conventional vehicles with passive functional subsystems are limited by their construction and cannot be changed in the course of motion. The driver, by his manual commands, gives the control quantities such as motion course (by steering system), vehicle rate (by driving and braking system), gradient of acceleration/deceleration (by driving/braking system), etc. The passive suspension system is distributed over all four tires of the vehicle. It consists of a spring damper and viscous shock absorber, i.e. the damper of a constant damping coefficient. The role of this system is to diminish the effects of unevenness of the road surface on the vehicle body and reduce inertia effects on the vehicle body caused by the changes in acceleration. Conventional vehicles[2] do not possess the subsystems that could actively influence dynamic behavior of the vehicle in the course of its motion. The dynamic behavior of the vehicles equipped with passive subsystems is determined by technical characteristics of

---

[2]Under the notion of conventional vehicles we understand the vehicles having only passive functional elements in their construction. They are presently still prevailing on the roads.

their construction and by the driver's skill. In contrast to conventional vehicles, the new-generation vehicles can be equipped with various active control systems such as active suspension system, active steering system, active drive system, Antilock Braking System (ABS), system of active tire shaft geometry [21], etc.

Active suspension system: Active suspension system was the first of active systems that was industrially built-in with conventional vehicles. It represents the vehicle control system intended for the realization of better dynamic performance of the vehicle body in the plane perpendicular to the road surface. The application of ASS results in a more uniform load of tire pneumatics and suspension joints and thus in smaller force amplitudes on the vehicle body in the vertical direction. This directly influences the quality of motion and ride comfort, but also the motion stability. With passive systems, the driver has no possibility to directly influence the "vertical" displacements of the vehicle body in the yaw, roll and pitch directions. By activating the vehicle suspension system it is possible to reduce the variation amplitudes of these load forces. In this way, a more uniform distribution of forces and better global dynamic behavior of the system are obtained. There are various types of active vehicle suspension systems.

Active traction/braking system: This important system is used in order to automatically regulate the angular velocity of the vehicle wheels. This system essentially improves the maneuvering capacity and reduces fuel consumption. The possibility of microprocessor control of the angular velocity of one wheel independently of the rotation speeds of the others, ensures better controllability of the vehicle body in the longitudinal and lateral directions and in the direction of yaw angle. This system offers the possibility of more efficient control of the longitudinal vehicle velocity, longitudinal position of the vehicle moving in a platoon, as well as of lateral position of the vehicle on the road. With conventional cars, the lateral position is controlled solely by the steering wheel, by giving the motion course of the front wheels.

Active steering system: This system represents an active system implemented for automatic changing of the tire - ground steering angle. Usually, it uses an electrohydraulic actuators to change the orientation angles of the wheels around their vertical axes. The possibility of independent change of the tire orientation angle yields better maneuvering abilities of the vehicle and ensures its better lateral controllability on the road. By applying the appropriate active system of changing tire orientation only on rear pair of wheels, it is possible to correct unskillful or inappropriate commands given by the driver. At the same time, this yields a better safety of motion, too.

1.7 Topics of the Monograph

The monograph consists of several topics that have been organized in chapters. The sequence of the chapters follows a logical order of exposition, so that the current consideration always relies upon the previously presented subject matter. In this way, we achieved a desired clarity of presentation and logical chain of thought from defining basic notions, presenting the known results from the actual research area, working out theoretical problems, verifying the results by simulation, to realizing the ultimate research goal. The monograph consists of the following topics:

- Introduction

- Modeling of Vehicle Dynamics

- Control of Automated Road Vehicles

- Stability Conditions of Road Vehicles

- Summary and Conclusions

At the beginning of each chapter there is a short abstract informing the reader about the contents of the current chapter. Chapter 2, dealing with modeling, contains a rather comprehensive survey of the results in the given research area, together with the concrete expressions, so that this chapter can be used as a handbook of its kind in road vehicle modeling. Chapter 3 describes a novel concept of the so-called integrated centralized control of automated road vehicle, representing an original solution of the authors. Chapter 4 presents also an original contribution to the analysis of vehicle stability on the basis of relatively complex spatial dynamic model of the system. The most important theoretical results are accompanied by the corresponding simulation examples, which is an essential contribution to the clarity of the presented subject matter and to the aspects of its practical implementation.

Bibliography

[1] Furukawa, Y., N. Yuhara, S. Sano, H. Takeda and Y. Matsushita, "A Review of Four-Wheel-Steering Studies from the Viewpoint of Vehicle Dynamics and Control", *Vehicle System Dynamics*, 18, pp. 151-186, 1989

[2] Ackermann J., Sienel W., "Robust Yaw Damping of Cars with front and Rear Wheel Steering", *IEEE Transactions on Control Systems Technology*, Vol. 1, No. 1, pp. 15-20, 1993

[3] Hirano Y., Ono, H., "Non-Linear Robust Control for an Integrated System of 4WS and 4WD", *Proceedings of the International Symposium on Advanced Vehicle Control AVEC '94*, Tsukuba, Japan, pp. 147-152, 1994

[4] Wakamatsu K., Akuta Y., Ikegaya M., Asanuma N., "Adaptive Yaw Rate Feedback 4WS with Friction Coefficient Estimator Between Tire and Road Surface", *Proceedings of the International Symposium on Advanced Vehicle Control AVEC '96*, Germany, Aachen, Vol. 1, pp.333-344, 1996

[5] Hirano, Y., "Development of the Integrated System of 4WS and 4WD by Control", *SAE Paper 930267*, 1993

[6] Bolzern P., Casati A., Locatelli A., Maroni M., "Control of Four-Wheel Steering Vehicles: A Design Based on μ-Synthesis Techniques", *Proceedings of the 13th Triennial World Congress*, San Francisco, USA, pp. 53-58, 1996

[7] Milliken Jr., W. F., "The Static Directional Stability and Control of the Vehicle", *SAE Paper 760712*, 1989

[8] Inagaki, Sh., Kshiro, J., Yamamoto, M., "Analysis on Vehicle Stability in Critical Cornering Using Phase-Plane Method", *Proceedings of the Int. Symposium on Advanced Vehicle Control AVEC'94*, Tsukuba, Japan, pp. 287-292, 1994

[9] Harada H., Araki Y., Oya M., "Control Effects of Active Rear Wheel Steering on Driver-Vehicle System", *Proceedings of the Int. Symposium on Advanced Vehicle Control AVEC'96*, Germany, Aachen, pp. 281-294, 1996

[10] Tagawa T., Morita K., Nagai M, "A Robust Active Front Wheel Steering System Considering the Limits of Vehicle Lateral Force", *Proceedings of the*

Int. Symposium on Advanced Vehicle Control AVEC'96, Germany, Aachen, pp. 315-332, 1996

[11] Yuhara N., Horiuchi S., Iijima S., "Improvement of Vehicle Handling Quality by means of Kinestethic Information", *Proc. of IEA '94*, Vol. 5, 1994

[12] Ackermann J., T., Sienel W., Jeebe H., Naab K., "Driving Safety by Robust Steering Control", *Proceedings of the Int. Symposium on Advanced Vehicle Control AVEC'96*, Germany, Aachen, pp. 377-394, 1996

[13] Nagai M., Yamanaka S., "Integrated Control Law of Active Rear Wheel Steering and Direct Yaw Moment Control", *Proceedings of the Int. Symposium on Advanced Vehicle Control AVEC'96*, Germany, Aachen, pp. 451-464, 1996

[14] Furukawa, Y., Abe, M., "On-Board-Tire-Model Reference Control for Co-operation of 4WS and Direct Yaw Moment Control for Imroving Active Safety of Vehicle Handling", *Proceedings of the Int. Symposium on Advanced Vehicle Control AVEC'96*, Germany, Aachen, pp. 507-526, 1996

[15] Wallentowitz, H., Longitudinal Control and Interaction with Other Systems, Contribution in Smart Vehicles, *Swets & Zeitlinger Publishers*, Netherlands, 1995

[16] Pruckner A., Seemann M., "Analysis of Dynamic Driving Control (DDC) Systems on Full Vehicle Model in ADAMS", *12th ADAMS user conference*, Marburg, Germany, 1997

[17] Rodić, A., D., Vukobratović, M., K., "Contribution to the Integrated Control Synthesis of Road Vehicles", *IEEE Transactions on Control Systems Technology*, Vol. 7, No. 1, pp. 64-78, 1999

[18] Rodić, A., Vukobratović, M., "Design of an Integrated Active Control System of Road Vehicles", *Journal of Computer Application in Technology, Special Issue on Active Structures*, Vol. 13, Nos. 1/2, pp. 78-92, January 2000

[19] Ono E., Yamashita M., Yamauchi Y., "Distributed Hierarchy Control of Active Suspension System", *Proceedings of the AVEC'94*, Tsukuba, Japan, pp. 373-378, 1994

[20] Vukobratović, M., Ekalo, Y., "New Approach to Control of Robotic Manipulators Interacting with Dynamic Environment", *Robotica*, Vol. 14, No. 1, pp. 31-39, 1995

[21] Potkonjak, V., Rodić, A., Vukobratović, M., "Roll Motion Control of Road Vehicles Using Tire Shaft Variable Geometry", *Proceedings of the 3rd ECPD International Conference on Advanced Robotics, Intelligent Automation and Active Systems*, Bremen, Germay, pp. 517-523, September 15-17, 1997

Chapter 2

Modeling of Vehicle Dynamics

Abstract

This chapter presents a complete survey of planar (simplified) and spatial (complex) vehicle models that can be found in the abundant literature. Special attention is paid to the nonlinear spatial model of the vehicle as an object belonging to the class of the so-called rigid-body dynamic systems. Such vehicle model, having 22 degrees of freedoms of motion, is used in the subsequent chapters in the dynamic controller synthesis and stability analysis. This model represents a compromise between the criterion of numerical complexity which is important for practical implementation in a real system, and the criterion of accuracy of describing dynamic behavior of that real system. The chapter is concerned with the following simplified vehicle models: quarter-car model, half-car planar model, single track model, planar model of the longitudinal and lateral dynamics, etc. Also, a relatively detailed description is given of the tire model as the vehicle element causing most of the nonlinearities in the system. We also present mathematical models of the particular active control subsystems such as: the active suspension system, the system of active driving/braking, and the active steering system. The choice of the appropriate types of actuators suitable for the application within the active subsystems is given along with their general mathematical model. Finally, the so-called integrated vehicle model, encompassing all the real system elements needed for the controller synthesis is presented. For the purpose of vehicle model simulation, velues of the parameters were taken from the available literature. The chapter ends with the description of some characteristic simulation examples based on the spatial model, reflecting the dynamic performance of the considered system. The Appendix provides a list of the parameters used in the simulation experiments.

2.1 Introduction

Recent advances in computer technology have been accompanied by advancement in the development of software tools for modeling, simulation, design, and control of large-scale dynamic systems. The design needs have imposed the requirements for the highest possible degree of accuracy of modeling the overall dynamics of vehicles, as well as their mechanical, hydraulic, and other subsystems. This yielded the development of specialized dedicated software packages for modeling and optimization of design characteristics and control of road vehicles.

Numerical simulation in the field of vehicle dynamics has made essential advancements in the last decade, with a remarkable progress in the mechanical analysis programs [1]. Owing to this progress, multi-body systems in a vehicle such as suspension systems, hydraulic power-steering system, elastic knuckle, etc. have been effectively modeled. A remarkable feature of the vehicle dynamics modeling is that the object "structure" is described instead of the "character" of the object behavior. A structure description model requires design information as its input (e.g. CAD model), and it involves all characteristics inherent in the structure itself, while a character description model reveals only the specified characteristics. Therefore, in the automobile industries, holding the overall vehicle design information, it has become common to apply mechanical analysis programs to vehicle dynamics analysis.

Building up a full vehicle model for mechanical analysis program requires detailed design information and various modeling techniques. To reduce such complicated tasks to a manageable job, Digital Vehicle System (DVS) was developed in 1986 [1], with ADAMS [2] as its solver. DVS contains a macro-system with hierarchical structure, which is a kind of knowledge database. Each macro consists of procedures to build up a sub-structure of a vehicle model. Also, the use of this software enables precise modeling of tire force, elastic structure modes, actuators' delay, and the like [1]. Apart from the mentioned DVS package, dedicated to modeling of road vehicles, we should also mention packages of similar dedication, like SIMPACK and ADNECS [3], which have been successfully applied in the DLR laboratories in Germany. Modern simulation programming package BAMMS, developed in the TNO-Road-Vehicles Research Institute, the Netherlands [4], is intended for geometrical and dynamical simulation of the overall vehicle dynamics. This simulation tool allows the user to create both simple vehicle models using motion equations and complex object models, using simulation tools for modeling of multi-body dynamic systems [4].

As mentioned above, there are nowadays a relatively large number of specific software packages for professional modeling of complex dynamic objects such as road vehicles. In this research monograph we will not deal with the problems of modern software tools intended for modeling and simulation. We shall only note that significant advances have been made in this field and that the existing software, possessing also the appropriate real-time graphical interface, enables simulation of technical systems of arbitrary complexity with a high speed and accuracy. As the ultimate goal of our research described in this monograph is the synthesis of automatic control of road vehicles, we shall deal first with vehicle modeling to such an extent which is needed for the synthesis of dynamic control algorithms. In view of this, conventional techniques of dynamic systems modeling, based on deriving differential equations of the object's behavior, are of interest for our further research and consideration. We shall give below a survey of the most known models of road vehicles that have been used in the synthesis of various algorithms for automatic control. We start with structurally most simple models and go towards more complex ones. Our ultimate goal is to present a new, relatively complex, spatial nonlinear vehicle model which simultaneously satisfies the accuracy criteria of describing real behavior of the system considered and numerical criterion of moderate numerical complexity, suitable for real-time implementation.

From the point of view of mechanics, a vehicle represents a complex, distinctly nonlinear, dynamic system with a relatively large number of degrees of freedom (DOFs), belonging to the class of large-scale dynamic systems (Fig. 2.1). These DOFs (independent motion directions), especially those that essentially influence the system's stability and quality of its dynamic behavior, have to be controlled. The vehicle stability, quality of its dynamic behavior and maneuvering capabilities depend to a great extent on the structure of the system and performances of its power-actuating subsystems. Selection of a best control strategy and its realization, within the limits of technical capacities of the system, represents a delicate problem whose solving requires knowledge of dynamic behavior of the vehicle under different conditions of motion. To that end, mathematical model of the vehicle dynamics in the interaction with its dynamic environment[1] represents a starting point in the stage of system design and its automatic control.

In the abundant literature from this field one can find various dynamic models of road vehicles. The majority of the known models are significantly simplified as they describe the vehicle dynamics in particular motion directions. Thus the vehicle models are reduced to the models describing real systems having only several DOFs in particular planes (plane of the road surface or the plane perpendicular to it). Such a pragmatic approach to modeling has certain advantages in practice. Many techniques known from control of linear systems are based on the application of simplified linear or linearized models. On the other hand, simplified models are of limited value as they are valid only in a relatively narrow range of variation of control variables and state variables. These limitations influence also the quality of the synthesized control and desired system stability. Having this in mind, this chapter will be concerned with most frequently used dynamic models of road vehicles and their important functional elements such as suspension system (SS) and tires. Besides, a complex spatial road vehicle model, suitable for the synthesis of dynamic control laws and stability analysis of the complete vehicle, will be described in detail.

The vehicle dynamics models can be categorized in two main groups: (i) pla-

---

[1]The notion of the dynamic environment of a road vehicle was defined in the introductory chapter.

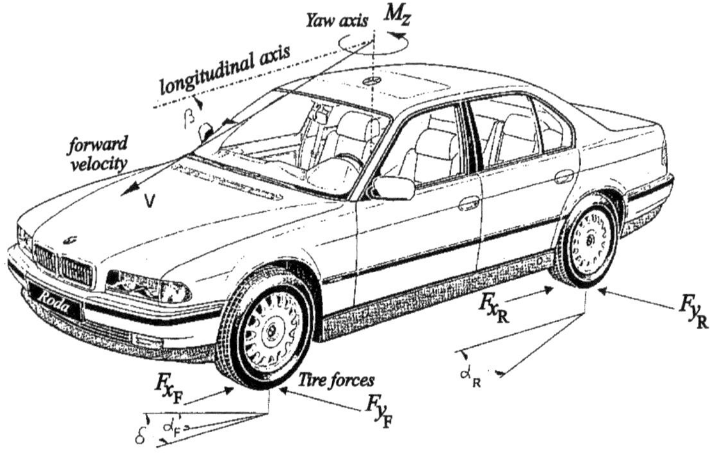

Figure 2.1: Contemporary road vehicle [5] as a complex dynamic system

nar models, and (ii) spatial models. Planar models represent simplified object models. They describe the system's dynamics in a reduced number of coordinates of motion directions, i.e. they deal with the dynamics of particular motion DOFs. None of these models describes the overall dynamic behavior of the vehicle simultaneously in all of motion directions. Planar models describe the so-called "planar dynamics" of the object, i.e. the vehicle dynamics in certain planes of the system displacement. To the group of planar models belong the known models such as "quarter-car vehicle model", "half-car vehicle model", "single track" or "bicycle" vehicle model, as well as "planar model of longitudinal and lateral dynamics". In contrast to planar vehicle models, spatial models describe the object's behavior in all six main directions of motion: three translational and three rotational. Translational directions of vehicle motion represent the directions of linear displacement of the vehicle body mass center (MC) in the longitudinal, lateral and (conditionally) vertical[2] directions. Rotational motion of the vehicle body represents angular displacements of the vehicle body about its main inertia axes: roll[3], pitch[4], and yaw[5] axes. Depending on the degree of its complexity, a spatial dynamic vehicle model may take also into

---

[2]We talk of the direction perpendicular to the road surface plane.

[3]The direction of this axis coincides with the longitudinal symmetry axis of the vehicle body.

[4]This axis is perpendicular to the longitudinal symmetry axis which is at the same time parallel with the road surface plane.

[5]This axis passes through the vehicle body MC and is perpendicular to the road plane.

account dynamic behavior of the other DOFs of motion that are not included in those six mentioned above. In this way it is possible to take into account both the elastic modes of the vehicle structure and local displacements at particular joints of the mechanism SS for each vehicle tire. Thus a spatial model may acquire additional complexity, which enhances its exactness but also, increases its numerical complexity. In the text below we give a survey of the known models of road vehicles that have been most often used in the research carried out in this field.

2.2 Planar Vehicle Models

The mentioned simplified planar models of road vehicles most frequently used in the literature are:
   (i) quarter-car vehicle model,
   (ii) half-car vehicle model,
   (iii) single track (or "bicycle") model of road vehicle and,
   (iv) model of longitudinal and lateral vehicle dynamics.

2.2.1 Quarter-car Vehicle Model

The quarter-car vehicle model represents a comparatively simplest model of vehicle dynamics. It is in fact a mathematical model of the vehicle SS, whose functional-symbolic scheme is presented in Fig. 2.2. From the viewpoint of mechanics, a road vehicle can be considered as a rigid body moving on its four wheels, interacting thus with the road surface. Under the notion of "vehicle body" is understood its structure in which is concentrated the system's mass. The vehicle body can be considered as the object structure defined by its geometry and dynamic parameters (mass, moments of inertia, coefficients of aerodynamic resistance, etc.). The structure is supported at four points of its elastic "supports" to which is connected by the joints (see Fig. 2.2). The mentioned "supports" represent the vehicle SS.

The role of SS of a road vehicle is manifold. First of all, the system ensures the vehicle's support on the ground. Further, it serves to damp the impact action of the road surface irregularities during the vehicle motion. Also, SS transmits the driving forces from the wheels onto the vehicle body, enabling thus its motion. Performances of the SS and tire pneumatics have a dominant influence on the vehicle behavior during motion. They give to it a nonlinear character of behavior, which additionally complicates the model of the overall system and its implementation in control. The adjustment of parameters of the active suspension system (ASS) offers the possibility of active control of dynamic performance of the vehicle in the direction perpendicular to the road surface. In this way is possible to directly improve performance of the vehicle stability, driving safety, and riding comfort.

Quarter-car vehicle model describes dynamic behavior of the vehicle in the direction perpendicular to the road surface plane. This model usually describes

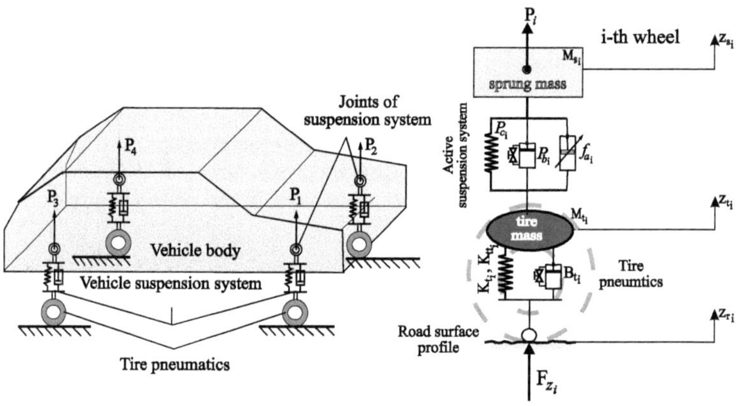

Figure 2.2: Quarter-car vehicle model - Functional-symbolic scheme of the vehicle suspension system

up to two DOFs of motion (see Fig. 2.2). It is assumed that the whole vehicle mass is reduced to the tires in the form of a weight whose mass is called "sprung mass". In Fig. 2.2 the sprung mass is represented by a weight which, via a system of springs and viscous dampers, is supported on the road surface. The considered SS model represents a planar model as it takes into account only the vehicle dynamics in the vertical plane of tire SS. In the same plane are also given the changes in the road surface profile. The force P_i in Fig. 2.2 represents the vertical reaction force at the i-th SS joint, i.e. payload; force F_{z_i} is the tire load, i.e. the adhesion force of the i-th tire onto the ground; P_{c_i} is the spring force of the spring damper to damp the impact-pulse action on the vehicle body; P_{b_i} is the damping force at the viscous shock absorber; f_{a_i} is the active damping force in the active damping cylinder; K_{t_i} is the stiffness coefficient of the tire pneumatic; B_{t_i} is the damping coefficient of the tire pneumatic; M_{s_i} and M_{t_i} are the SS sprung mass and tire mass, respectively.

2.2.1.1 nonlinear Model of the Suspension System

Dynamic model of the tire ASS shown in Fig. 2.2 is a nonlinear function of the local[6] coordinates of state. The nonlinear SS dynamic model can be described by the following pair of scalar differential equations.

$$
\begin{aligned}
M_s \ddot{z}_s + P_b + P_c &= -M_s g + f_a \\
M_t \ddot{z}_t + T_b + T_c &= -M_t g + P_b + P_c - f_a
\end{aligned}
\tag{2.1}
$$

---

[6]Local state coordinates represent the corresponding positions and velocities of the characteristic points of the SS shown in Fig. 2.2.

The stiffness (elasticity) forces P_c, T_c and the damping forces P_b, T_b of the SS and tire pneumatic can be described by the following expressions:

$$P_b = \begin{cases} B_{s_1}(\dot{z}_s - \dot{z}_t) & |(\dot{z}_s - \dot{z}_t)| < \overline{w} \\ B_{s_1}\overline{w} + \overline{B}_{s_2}((\dot{z}_s - \dot{z}_t) - \overline{w}) & (\dot{z}_s - \dot{z}_t) \geq \overline{w} \\ -B_{s_1}\overline{w} + \overline{B}_{s_2}((\dot{z}_s - \dot{z}_t) + \overline{w}) & (\dot{z}_s - \dot{z}_t) \leq -\overline{w} \end{cases} \tag{2.2}$$

$$P_c = K_{s_1}((z_s - z_t) + K_{s_2}(z_s - z_t)^5) \tag{2.3}$$

$$T_b = B_t(\dot{z}_t - \dot{z}_r) \tag{2.4}$$

$$T_c = K_t((z_t - z_r) + K_{tt}(z_t - z_r)^3) \tag{2.5}$$

Here, use is made of the following notations[7]: M_s is the vehicle sprung mass $[kg]$; B_{s_1} and B_{s_2} are the damping coefficients of the SS viscous dampers $[N/(m/s)]$; K_{s_1} and K_{s_2} are the constants of the SS damping spring expressed in $[N/m]$ and $[m^{-4}]$; M_t is the tire mass $[kg]$; B_t is the damping coefficient of tire pneumatic deformation $[N/(m/s)]$; K_t and K_{tt} are elasticity constants of tire pneumatics $[N/m]$ and $[m^{-2}]$, and $\overline{w}$ is the maximal speed of the piston in the cylinder of the viscous damper.

At the SS joint[8], as well as on every tire pneumatic, the corresponding forces of vertical load P and F_z (see Fig. 2.2) are acting. The action of the force P at the suspension joints produces the corresponding displacement z_s of the sprung mass M_s. In this way, the orientation (position) of the vehicle with respect to the road surface is changed in an indirect way. The force F_z represents the vertical loading force of the tire pneumatic due to the payload. This force represents at the same time the force of the tire adherence to the ground surface. The amplitude of this force determines the forces of rolling and sliding friction of the wheel on the road surface, as well as the values of longitudinal and lateral forces of tire-road interaction. Intensities of the forces P and F_z are calculated from the equation of the mechanism equilibrium:

$$\begin{aligned} P &= M_s\,\ddot{z}_s + M_s\,g \\ P &= P_c + P_b - f_a \\ F_z &= T_c + T_b \end{aligned} \tag{2.6}$$

The state of rest represents a stable state of the system illustrated in Fig. 2.2, and it is characterized by the corresponding values of positions of the particular mechanism points and the values of forces at fictitious cross-sections of the mechanism. Equilibrium state is defined by the following values of state variables:

$$\ddot{z}_r = 0, \quad \dot{z}_r = 0, \quad z_r = 0,$$

---

[7]For simplicity, in the relations (2.1)-(2.5) the index $"i = 1,\ldots,4"$, used to denote state variables in Fig. 2.2, is ommited.

[8]The SS joint in Fig. 2.2 is represented by the point at which the tire SS is attached to the vehicle structure. The tire forces are transmitted to the SS joints, realizing thus their action on the vehicle body and bringing about its motion.

$$\ddot{z}_t = 0, \ \dot{z}_t = 0,$$
$$\ddot{z}_s = 0, \ \dot{z}_s = 0,$$
$$P = -M_s \, g$$
$$F_z = -(M_s + M_t) \, g \tag{2.7}$$

The SS deflection $e = z_s - z_t$ and deformation of tire pneumatics $e_t = z_t - z_r$ of the unexcited system in equilibrium are determined by solving the equations to follow, whereby the values for P and F_z from (2.6) are taken into account. Then

$$
\begin{aligned}
F_z &= T_c = K_t((z_t - z_r) + K_{tt}(z_t - z_r)^3) \\
P &= P_c = K_{s_1}((z_s - z_t) + K_{s_2}(z_s - z_t)^5)
\end{aligned}
\tag{2.8}
$$

The amplitude of the SS joint center position z_s and the tire mass center z_t at rest are determined after solving the equation (2.8), taking into account the kinematic relations:

$$
\begin{aligned}
z_t &= e_t + z_r \\
z_s &= e + z_t
\end{aligned}
\tag{2.9}
$$

The system state variables at the mechanism's state of rest are taken as the initial states in the model simulation.

2.2.1.2 Linear Model of the Suspension System

Linear model of the SS is frequently used because of its form which allows the use of conventional techniques known from the theory of automatic control of linear systems. By its structure, linear SS model is simpler than the corresponding nonlinear model. This is advantageous in some cases, but also has a certain negative consequences in respect of accuracy concerning the application of these models in control. In the case when simulation results for a linear and nonlinear model show small differences in the system's behavior for the same span of changes in the state coordinates and control variables, it is possible to use linear model for control purposes. Otherwise, it is not possible to guarantee satisfactory control accuracy and desired stability of the system. Experience shows that linear SS model is valid in case of relatively small amplitudes of deflection of the SS and tire pneumatics from their working points. In that case, the nonlinear nature of the system does not essentially come to the expression. Apart from practical application in the control synthesis, linear SS model can also be used for estimation of the approximate frequency characteristic of the system. This characteristic is used in the initial stages of design as well as in the determination of control gains of the ASS local controller. Otherwise, the system's frequency analysis is in the final instance carried out by experiment on a real system.

Linear dynamic model of SS, called quarter-car vehicle model or model of vertical dynamics, is described in detail in [6, 7]. The vehicle SS is modeled as an oscillatory mechanism whose parameters are: the sprung mass M_s, damping coefficient B_s, and stiffness coefficient K_s. It is assumed that the values of these quantities are constant. The other end of the SS is supported on the tire, which also possesses the corresponding physical characteristics. The damping coefficient of the tire pneumatic B_t is often omitted from the model as it is assumed that the damping force is negligibly small compared with the elastic characteristic of the pneumatic determined by the value of the stiffness coefficient K_t. Nonlinear ASS model defined by the expressions (2.1)-(2.5) can be simplified and represented in a linear form:

$$
\begin{aligned}
M_s \ddot{z}_s + B_s(\dot{z}_s - \dot{z}_t) + K_s(z_s - z_t) &= f_a \\
M_t \ddot{z}_t + B_t(\dot{z}_t - \dot{z}_r) + K_t(z_t - z_r) &= -f_a + B_s(\dot{z}_s - \dot{z}_t) + \\
&\quad + K_s(z_s - z_t)
\end{aligned}
\tag{2.10}
$$

In the model (2.10) are omitted the expressions for the gravity forces $M_s \cdot g$ and $M_t \cdot g$, appearing in the nonlinear model (2.1). These terms are omitted with the aim of deriving the transfer function of the linear SS model. Under assumption that the mentioned gravity forces are constant in the course of motion, their influence can be omitted from the model. Because of their constancy, the gravity forces produce constant deformations of the system's elastic elements. The effect of these forces on the SS and tire pneumatics is taken into account by determining initial deformations of various amplitudes, differing from the corresponding deformations determined by applying the nonlinear model (2.1)-(2.5). Apart from a static action of the mentioned gravity forces, inertial forces, damping forces, and spring elastic forces influence the mechanism's motion. For the sake of brevity in describing dynamic system's behavior, it is convenient to adopt the following local state coordinates:

$$
\begin{aligned}
x_1 &= z_s - z_t && \text{suspension system deflection} \\
x_2 &= \dot{z}_s && \text{sprung mass displacement velocity} \\
x_3 &= z_t - z_r && \text{tire pneumatics deflection} \\
x_4 &= \dot{z}_t && \text{tire bobbing velocity}
\end{aligned}
$$

The linear model of ASS (2.10) can be expressed in a vector form in the state space. The equation of the system state is of the form [6, 7]:

$$
\dot{x} = Ax + B\,u + F\,p
\tag{2.11}
$$

where x is a (4×1) state vector whose elements are $x = [x_1\ x_2\ x_3\ x_4]$; $u = f_a$ is the scalar quantity representing active damping force; $p - \dot{z}_r$ is the corresponding velocity which takes into account the character of variations in the road surface profile (Fig. 2.2) and represents an external disturbance exhibiting a destabilizing effect on the system. The matrices and vectors in the model (2.11) have the values:

$$
A = \begin{bmatrix}
0 & 1 & 0 & -1 \\
-\dfrac{K_s}{M_s} & -\dfrac{B_s}{M_s} & 0 & \dfrac{B_s}{M_s} \\
0 & 0 & 0 & 1 \\
\dfrac{K_s}{M_t} & \dfrac{B_s}{M_t} & -\dfrac{K_t}{M_t} & -\left(\dfrac{B_t}{M_t} + \dfrac{B_s}{M_t}\right)
\end{bmatrix};
$$

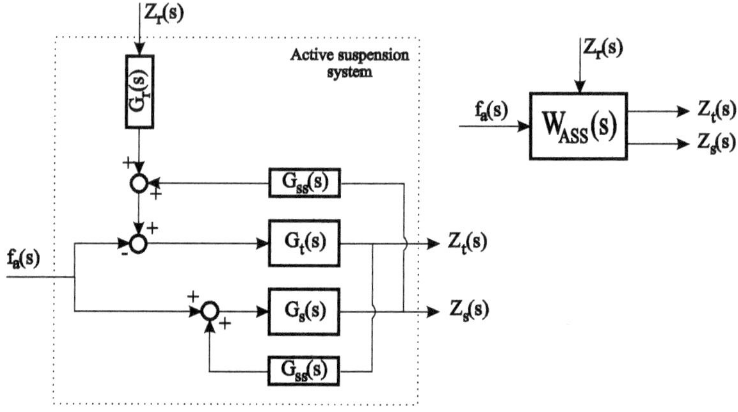

Figure 2.3: Block-diagram of the vehicle linear suspension system

$$
B \;=\; \begin{bmatrix} 0 \\ \frac{1}{M_s} \\ 0 \\ -\frac{1}{M_t} \end{bmatrix} ; \qquad
F = \begin{bmatrix} 0 \\ 0 \\ -1 \\ \frac{B_t}{M_t} \end{bmatrix} ; \tag{2.12}
$$

2.1.1.3 Transfer Function of the Suspension System

Starting from the differential equations of the behavior of SS given by the relations (2.10) and applying on them the Laplace transformation for all the initial conditions being equal to zero, it is possible to draw a block-diagram of the linear system. The resulting block-diagram is presented in Fig. 2.3. Its input control variable is represented by the signal $f_a(s)$; $Z_r(s)$ is the system disturbance signal which represents the Laplace transformation of the signal of road surface profile variation; $Z_s(s)$ and $Z_t(s)$ are output signals of the system, representing the corresponding SS displacements, as described above.

Transfer functions of the particular loops of the transfer system are:

$$
\begin{aligned}
G_r(s) &= B_t s + K_t \\
G_{ss}(s) &= B_s s + K_s \\
G_t(s) &= \frac{1}{M_t s^2 + (B_s + B_t)s + (K_s + K_t)} \\
G_s(s) &= \frac{1}{M_s s^2 + B_s s + K_s}
\end{aligned} \tag{2.13}
$$

Here, $W_{ASS}(s)$ denotes the ASS transfer function in its compact form.

2.2.1.4 Frequency Characteristics of the Suspension System

With the aim of determining optimal values of the parameters of the vehicle SS, as well as for the needs of adjustment (in the frequency domain) of control gains of the ASS controller, it is necessary to determine the values of resonance frequency and system bandwidth range. In this section we will show how frequency characteristics of the vehicle SS can be determined. The exact frequency characteristic of the nonlinear system can be determined only by experiment, by measuring the corresponding signal values on a real system [8]. The system shown in Fig. 2.3 by its block-diagram is excited externally from a state of rest (equilibrium state) by the action of a simple periodic excitation signal $z_r(t)$. Then, the signals of the object response are saved, i.e. the values of the output variable $z_s(t)$ are recorded over a certain time interval t until a stationary state is attained. The excitation signal is given in the form of a simple periodic signal whose shape and time derivative are given by the expressions:

$$
\begin{aligned}
z_r(t) &= a_1 \, sin(\omega t) \\
\dot{z}_r(t) &= a_1 \omega \, cos(\omega t) \\
\ddot{z}_r(t) &= -a_1 \omega^2 \, sin(\omega t)
\end{aligned}
\tag{2.14}
$$

In the procedure, the signals z_r of constant amplitude $a_1 = const$ and different eigenvalues of the angular frequency ω are successively given as input. The values of the assigned angular frequencies ω are taken arbitrarily from the set $\omega \in (0, \omega_{max}]$, where ω_{max} is the assumed maximal frequency value for the mechanism considered. On the basis of experience this frequency in case of mechanical systems is considered to be in the range up to $\omega_{max} \sim 60 \, [rad/s]$. After the system's excitation with a periodic signal, values of the output signal $z_s(t)$ are saved. The latter signal is also of the periodic form $z_s(t) = a_2(\omega) \cdot sin(\omega t + \psi(\omega))$. It differs from the excitation signal $z_r(t)$ in respect of its amplitude $a_2 \neq a_1$ and the phase delay $\psi(\omega)$. By comparing recordings of the signal $z_r(t)$ with $z_s(t)$ for the same values of frequency ω it is possible to determine the corresponding signal amplitude a_2 (a_1 is given as a constant value) and the phase delay of the one signal with respect to the other. On the basis of the measured amplitude values one can calculate the value of the module $M(\omega) = a_2/a_1$. The variable $M(\omega)$ is called the amplitude characteristic module of the system. The measured value of the phase delay $\psi(\omega)$ of one signal with respect to the other is called phase abundance of the system, and its value is characteristic for a particular frequency ω. Values of the variable quantities $M(\omega)$ and $\psi(\omega)$ are graphically presented in the form of the functional diagrams $\omega - M(\omega)$ and $\omega - \psi(\omega)$. These diagrams are called amplitude-frequency characteristics and phase-frequency characteristics of the system. On the basis of the shape of the mentioned curves in the diagrams it is possible to estimate the quality of the object's behavior.

Parameter	Unit	Value
M_s	$[kg]$	335.0564
B_{s_1}	$[\frac{N}{m}]$	10000
B_{s_2}	$[\frac{N}{m}]$	4000
K_{s_1}	$\frac{N}{m}$	40000
K_{s_2}	$[m^{-4}]$	40000
$\overline{w}$	$\frac{m}{s}$	0.025
M_t	$[kg]$	13
B_t	$[\frac{N}{m}]$	1916.5
K_t	$\frac{N}{m}$	196200
K_{tt}	$[m^{-2}]$	5020

Table 2.1: Dynamic parameters of the SS model used in the simulation example

2.2.1.5 Simulation Experiments with Model of Suspension System

Example 2.1: Characteristics of the vehicle SS are of great importance to stability of the overall vehicle. For this reason, in the design stage, it is necessary to pay special attention to the determination of optimal values of the SS parameters, to ensure a desired quality of dynamic behavior of the vehicle during motion. As has already been mentioned, the SS parameters must be determined so that the system is neither too "soft" nor too "hard", to attain a compromise between the criterion of riding comfort and the desired stability. To that effect, the results of model simulation offer useful information for the analysis of behavior of the designed SS. In this section we present some characteristic simulation results of both linear and nonlinear SS models (Fig. 2.2) for the case involving no active action of the viscous cylinder ($f_a = 0$).

Dynamic parameters of the SS models used in simulations were taken from [9]. Their values are given in Table 2.1. The initial conditions given in simulation were determined from the relations (2.7)-(2.9):

$$\ddot{z}_r = 0, \ \dot{z}_r = 0, \ z_r = 0,$$
$$\ddot{z}_t = 0, \ \dot{z}_t = 0, \ z_t = -0.0109 \, [m]$$
$$\ddot{z}_s = 0, \ \dot{z}_s = 0, \ z_s = -0.0683 \, [m]$$
$$e_t = -0.0109 \, [m]$$
$$e = -0.0574 \, [m]$$
$$P = M_s \cdot g = 3286.9 \, [N]$$
$$F_z = (M_s + M_t) \cdot g = 3414.4 \, [N] \tag{2.15}$$

For the given SS parameters, simulation of the model (2.1) was performed. The diagrams of the type *force-deformation* and *force-deformation rate* were

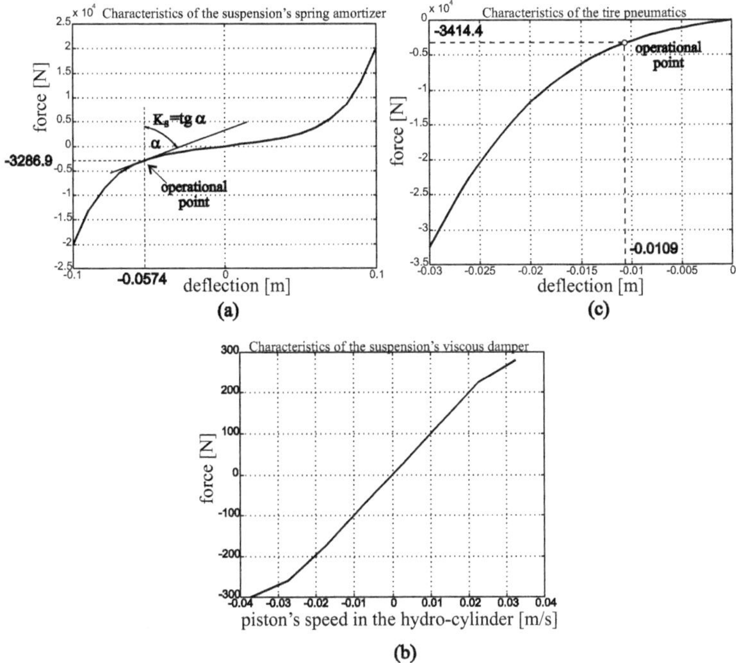

Figure 2.4: a) Characteristic of the SS spring amortizer, b) Characteristic of the SS viscous shock absorber, c) Characteristics of the tire pneumatic

constructed. In the graphs presented in Fig. 2.4 are shown: SS spring force, viscous damping force of the piston in the cylinder and, elastic deformation of the tire pneumatic. Coordinates of the so-called "operational points" shown in the diagrams correspond to the values of the forces and deformations that can be measured in the rest state of the SS or in the state corresponding to a steady-state motion of the vehicle at a constant speed on an ideally smooth (level) road surface. At the operational points, it is possible to draw tangents on the obtained curves of functional relationships. Slopes of these tangents determine the coefficients of stiffness (elasticity) K_s and K_t of the spring and tire pneumatic in the relations (2.10).

In the simulation experiment, behavior of both the linear and nonlinear models of the considered system is analyzed for the same system parameters and conditions of motion. For this purpose, the vehicle motion on an uneven road is given, and two cases are examined. In the first case is supposed that the tire hits a "sharp" hump on the road, approximated by a step function (Fig. 2.5a). In the other case we studied the tire motion on a road whose profile is a randomly changing quantity approximated by a random function with the given maximal amplitude (Fig. 2.5b). The objective of the simulation experiment was to compare behavior of the nonlinear and linear SS model, using the same perturbation signal. In Figs 2.5a and 2.5b are illustrated the corresponding simulation results of the above nonlinear (2.1) and linear (2.10) SS models. The excitation signals shown in Fig. 2.5 have a different effect on the system. In both cases the amplitude of the road surface unevenness does not exceed a maximal value $z_r^{max} = 10\ [mm]$. In the case of simulation of the linear model (2.10) the adopted model parameters were: $K_s = 5.7263 \times 10^4\ [N/m]$ and $K_t = 3.1325 \times 10^5\ [N/m]$, whose values were determined from the corresponding graphs shown in Fig. 2.4. In Figs 2.5a and 2.5b are given the responses of the nonlinear and linear SS models to the given excitation signals. On the basis of analysis of the obtained system responses it can be concluded that the linear model approximates relatively well the nonlinear behavior of the system to the moment when deformations of the SS and tire pneumatic in the course of motion do not deviate significantly from their values at the operational points. The more pronounced road roughness, the greater are differences in the behavior of the nonlinear and linear models. In the case of involvement of larger amplitudes of external forces acting on the system, i.e. in the case of the occurrence of larger deformations of the SS and tire pneumatics, the differences in behavior of the linear model (2.10) and nonlinear model (2.1) become more pronounced. This means that the linear model is a rather crude approximation of a real system. For this reason this model could not be used in the synthesis of dynamic control laws without introducing additional compensators to compensate for the structure inaccuracies of the model used.

Example 2.2:   The quality of behavior of the system and its stability can be judged on the basis of the shape of its frequency characteristic. Adopting the model parameters from Table 2.1 and using the developed SS model given by the relations (2.1)-(2.5), the SS frequency characteristic was determined on the basis of the results obtained in the following simulation experiment. The system was successively perturbed with the signal $z_r(t)$ whose periodic form is defined by the relation (2.14). In Fig. 2.6 are presented three pairs $(z_r,\ z_s)$ of input-output signals as a result of simulation of the nonlinear model described by the relations (2.1)-(2.5). Three different values of angular frequency were given, viz. $\omega_1 = 12.56\ [rad/s]$, $\omega_2 = 17.59\ [rad/s]$ and $\omega_3 = 25.13\ [rad/s]$. In the diagram (Fig. 2.6) one can notice the differences in the values of the amplitudes a_1 and a_2 of input and output signals, as well as the phase delay between the two signals. In Fig. 2.7 are presented the amplitude-frequency and phase-frequency characteristics of the SS illustrated in Fig. 2.2, whose parameters were taken from Table 2.1. By analyzing the frequency characteristic shown in Fig. 2.7 it is possible to perceive the maximal value of the module amplitude, the so-called "resonance peak" of the system. It is $M_r(\omega) = 2.398$. The frequency

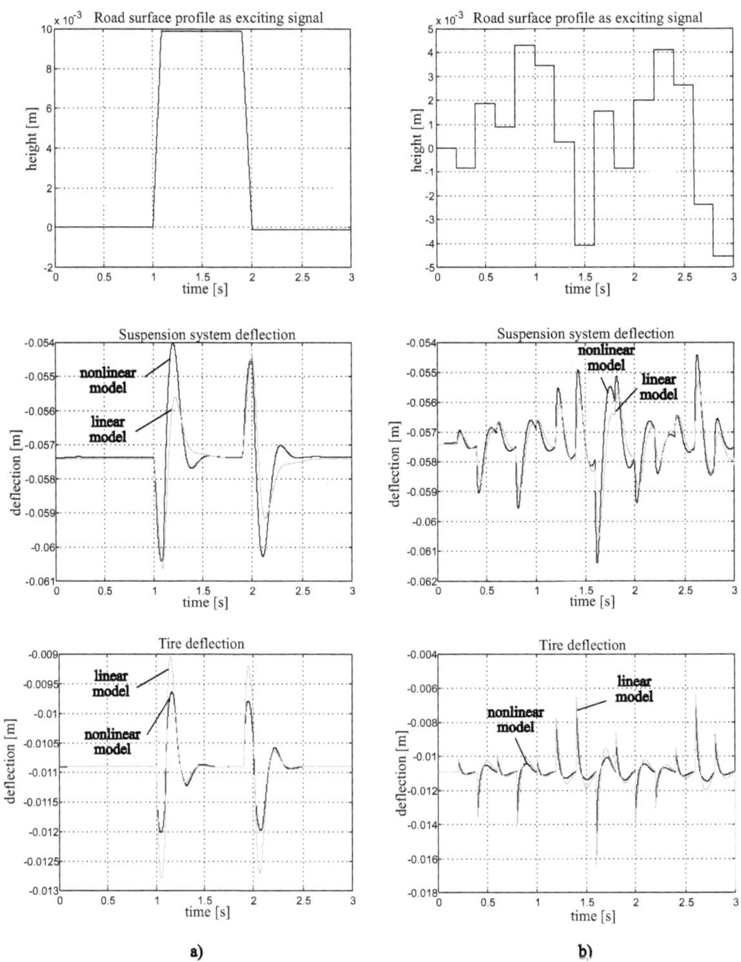

Figure 2.5: Characteristics of the behavior of the linear and nonlinear SS for different excitation signals: a) step function, b) random function

corresponding to this amplitude has a value $\omega = \omega_r = 17.59\ [rad/s]$. The system bandwidth range is determined from the graph, having in mind that at this frequency the system has an approximate module value $M(\omega) = 0.707$. The angular frequency corresponding to this value of amplitude gain has a value $\omega_n = 45.86\ [rad/s]$. As for the phase abundance, two characteristic values ω_r and ω_n are of interest. These are: $\psi_r(\omega_r) \sim -0.7716\ [rad]$ and $\psi_n(\omega_n) \sim -1.5401\ [rad]$. Excitation signals z_r of higher frequencies ($\omega > \omega_n$) have a relatively smaller effect on the system behavior. The z_r signals of high frequencies ($\omega >> \omega_n$) cause no displacement of the mechanical system but produce micro-vibrations in it and an unpleasant accompanying noise.

In this example, by analyzing the shape of the obtained frequency characteristics, the intention was to demonstrate the advantages offered by ASS in comparison with conventional, passive SS. With this aim we simulated behavior of the ASS possessing active viscous damper which generates an active force in the cylinder piston $f_a(t)$. At that, the possibility was given of a linear change of the damping coefficient $d(t)$ (Fig. 2.8a) in dependence of the current value of speed variation ($\dot{z}_s(t) - \dot{z}_s(t)$) of the viscous damper piston. In this way, additional compensation of deformations of the SS was attained by realizing active damping force according to the law $f_a(t) = -d(t) \cdot (\dot{z}_s(t) - \dot{z}_s(t))$. By doing this we determined again the frequency characteristic of ASS[9]. This is presented in Fig. 2.8b. By comparing the characteristic indicators from Fig. 2.7 and Fig. 2.8b it can be seen that the resonance peak M_r in case of applying ASS $M_r(\omega_r) = 2.076$ is smaller compared with the corresponding resonance peak when applying PSS ($M_r(\omega_r) = 2.398$). This obviously indicates that ASS can directly damp the system's oscillations. It should be mentioned that in the synthesis of the ASS controller one has to take into account that control gains in the chosen control law are determined (in frequency domain) so that their values in no case "correspond" to the frequency values that are close or equal to the resonance frequency of the system. Otherwise, they would cause undesired gains of the SS deflection amplitude. This could result in the vehicle instability during its motion, and even yield the damages of particular elements of the SS and tires as a consequence of excessive dynamic loads.

2.2.2   Half-Car Vehicle Model

Pitch and heave motions of the vehicle body represent the motion DOFs in the plane perpendicular on the road surface, and these motions exhibit strong dynamic coupling. Displacements of the vehicle body in these directions are the consequences of forces acting on tires. Variations of the load on each tire caused by inertial effects inside the system or by the impact of variation of the road surface profile on front and rear tires, have a significant effect on the vehicle dynamics behavior during motion. Half-car vehicle model has been developed with the aim of investigating the vehicle dynamics in the plane perpendicular to the road surface with respect to the longitudinal axis of the vehicle body.

---

[9]In Fig. 2.7 is presented the frequency characteristic of passive suspension system (PSS).

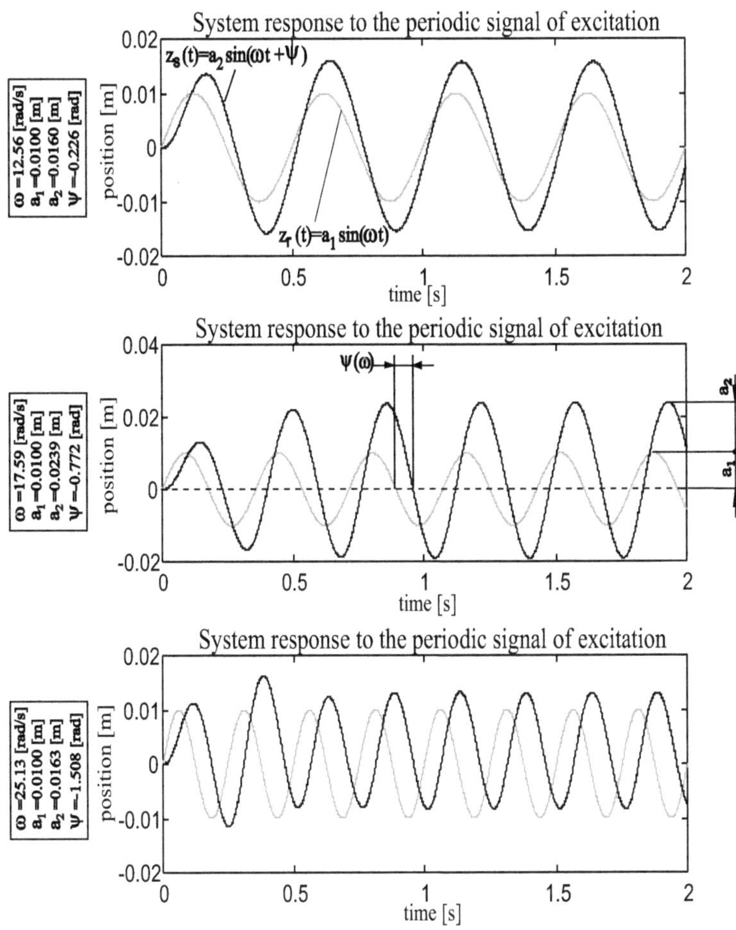

Figure 2.6: Simple periodic excitation signal and system response to the given signal for different oscillation frequences

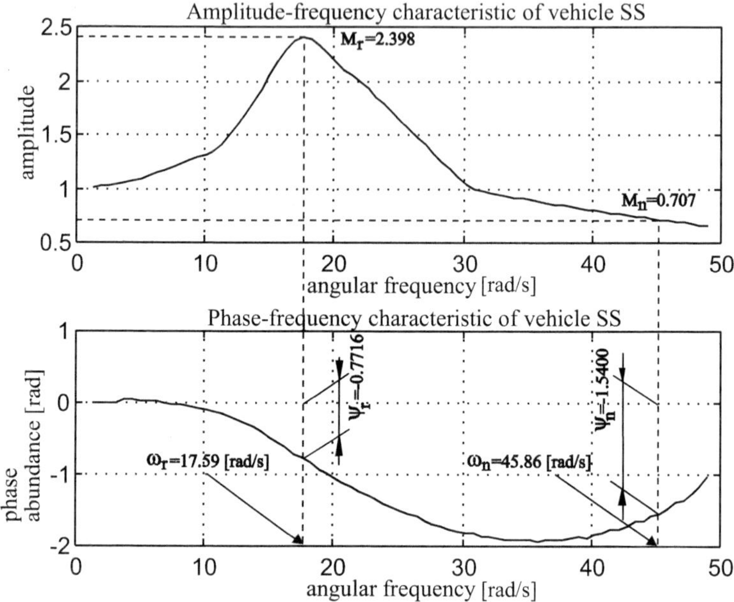

Figure 2.7: Amplitude-frequency and phase-frequency characteristics of the SS shown in Fig. 2.2

In comparison with the quarter-car vehicle model this model is more complete. Because of that this model describes more exactly the so-called "vertical dynamics" of the vehicle. Use of the half-car vehicle model allows an adequate description of the inertia effects of braking and acceleration during the vehicle motion.

In Fig. 2.9 [10] is presented the half-car vehicle model in the plane perpendicular to the plane of motion. This model describes the mechanism having six DOFs of motion: two on each of two wheels in the vertical direction of bobbing, one in the direction of heave motion of the vehicle MC, and one DOF in the direction of changing pitch angle of the vehicle body. In Fig. 2.10 [10] are presented in symbolic form all DOFs of the mechanism's motion.

In Figs 2.9a and 2.9b are denoted the reaction forces (F_{z_l}, F_{z_t}) of the SS acting on front (leading - with the index "l") and rear (tracking - with the index "t") wheel. The deformation forces of SS and of tire pneumatics on the leading and tracking wheels caused by longitudinal acceleration maneuvering are given

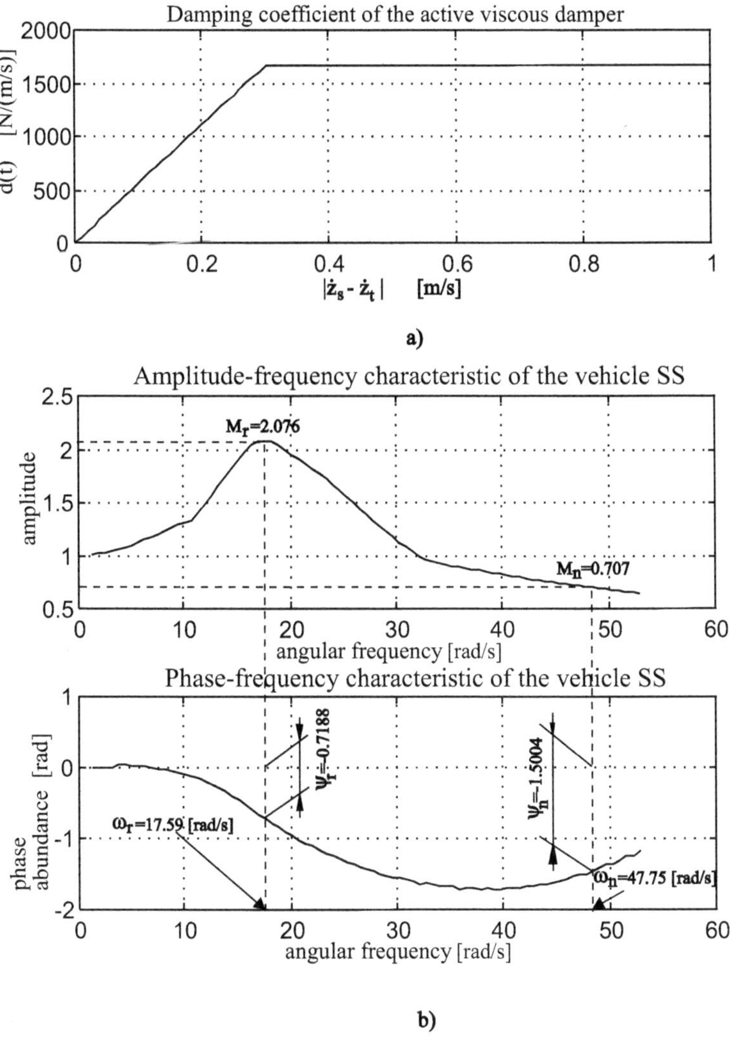

Figure 2.8: a) Law of change of damping coefficient of the active viscous damper of SS, b) Frequency characteristics of ASS

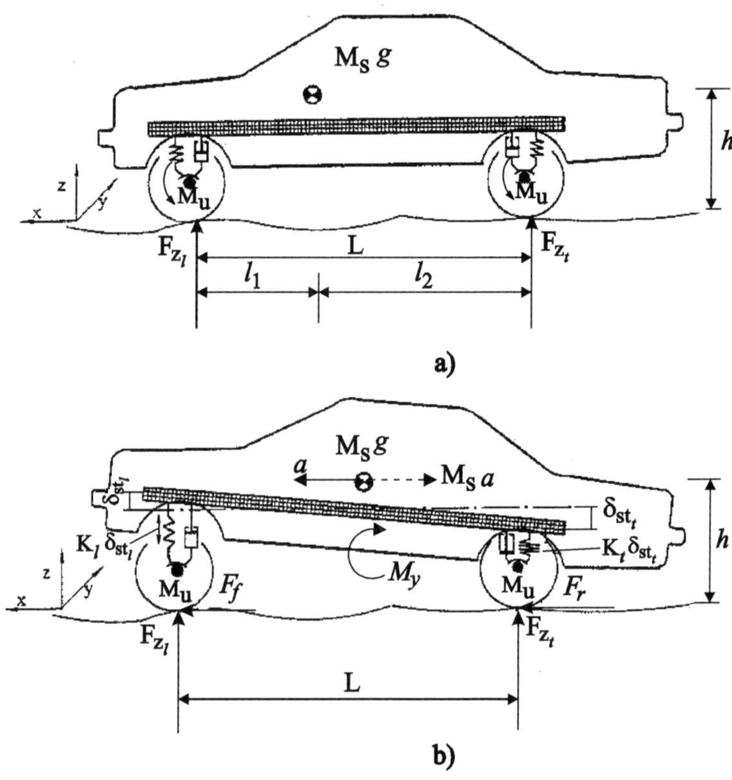

Figure 2.9: a) Half-car model of the vehicle in the state of rest or uniform motion in a straight line, b) Half-car vehicle model in the phase of vehicle acceleration

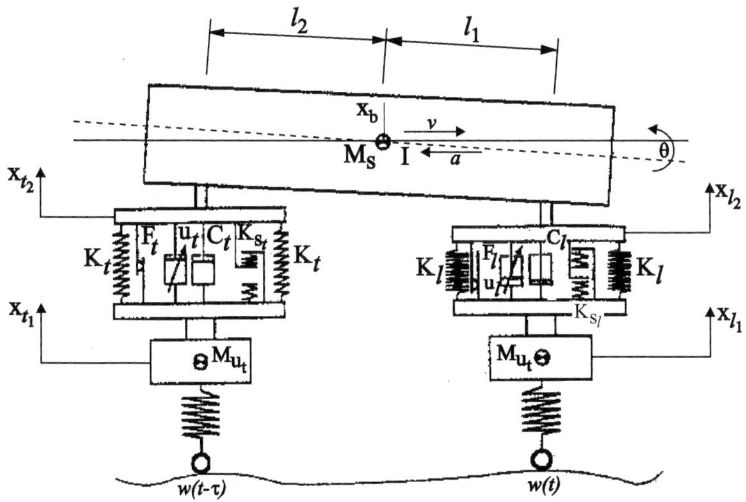

Figure 2.10: Symbolic scheme of the half-car vehicle model with ASS [10]

by the expressions [10]:

$$\frac{M_s\, a\, l_2}{L} = K_l\, \delta_{st_l} = K\delta_{t_l}$$

$$\frac{M_s\, a\, l_1}{L} = K_t\, \delta_{st_t} = K\delta_{t_t} \tag{2.16}$$

whereas total vertical tire-road reaction forces are described by:

$$F_{z_l} = M_s\, g\, \frac{l_2}{L} + M_u\, g - K_l\, \delta_{st_l}$$

$$F_{z_t} = M_s\, g\, \frac{l_1}{L} + M_u\, g + K_t\, \delta_{st_t} \tag{2.17}$$

In Fig. 2.10 is presented the half-car vehicle model with ASS, which has non-linear time-variable characteristics. The dynamic equations describing the "vertical" ascending/descending $x_b(t)$, pitching $\theta(t)$ about the axis passing through the vehicle MC, as well as bobbing of the sprung mass M_s and unsprung mass M_u on the front and rear tires are given by the expressions [10]:

$$\begin{aligned}
M_s(t)\ddot{x}_b = {} & u_l + u_t - F_{C_l} - F_{C_t} - F_{k_l} - F_{k_t} - \\
& - F_{d_l} - F_{d_t} - F_{s_l} - F_{s_t} + K_l\delta_{st_l} + K_t\delta_{st_t} - \\
& - M_s(t)\frac{l_1}{L}g - M_s(t)\frac{l_2}{L}g + M_s(t)\frac{h}{L}a - M_s(t)\frac{h}{L}a
\end{aligned}$$

$$I(t)\ddot{\theta} = (F_{C_t} + F_{k_t} + F_{d_t} + F_{s_t} + M_s(t)\frac{a\,h}{L} - K_t(t)\delta_{st_t} - u_t)l_2 -$$
$$- (F_{C_l} + F_{k_l} + F_{d_l} + F_{s_l} - M_s(t)\frac{a\,h}{L} - K_l(t)\delta_{st_l} - u_l)l_1$$

$$M_{u_l}(t)\ddot{x}_{l_1} = F_{C_l} + F_{k_l} - u_l + F_{d_l} + F_{s_l} -$$
$$- K_l(t)\delta_{st_l} - K(t)(x_{l_1} - \omega(t)) + K(t)\delta_{t_l}$$

$$M_{u_t}(t)\ddot{x}_{t_1} = F_{C_t} + F_{k_t} - u_t + F_{d_t} + F_{s_t} -$$
$$- K_t(t)\delta_{st_t} - K(t)(x_{t_1} - \omega(t - \tau)) + K(t)\delta_{t_t} \tag{2.18}$$

where

$$F_{C_i} = F_i(t)\,sgn(\dot{x}_{i_2} - \dot{x}_{i_1}), \quad \text{where the sign } i = "l" \text{ or } i = "t"$$
$$F_{k_i} = K_i(t)(x_{i_2} - x_{i_1})$$
$$F_{d_i} = C_i(t)|\dot{x}_{i_2} - \dot{x}_{i_1}|(\dot{x}_{i_2} - \dot{x}_{i_1})$$
$$F_{s_i} = K_{s_i}(t)S[x_{i_2} - x_{i_1} - \frac{\delta}{2}sgn(x_{i_2} - x_{i_1})]$$
$$S = \begin{cases} 0 & \text{if } |x_{i_2} - x_{i_1}| \le \frac{\delta}{2} \\ 1 & \text{in other cases} \end{cases} \tag{2.19}$$

The notations $M_s(t)$, $I(t)$, $M_{u_l}(t)$ and $M_{u_t}(t)$ represent the vehicle sprung mass, polar moment of inertia about the vehicle pitch axis passing through MC, and unsprung mass, i.e. the mass of the leading and tracking tire. Components of the forces F_{C_l}, F_{C_t}, F_{k_l}, F_{k_t}, F_{d_l},F_{d_t}, F_{s_l} and F_{s_t} represent the respective forces of Coulomb's friction on the tires, the SS stiffness forces of the front and rear tires, viscous damping forces in the viscous shock absorbers and elastic limiting forces of vehicle stopping. Models of these forces contain the following coefficients: F_l, F_t, K_l, K_t, C_l, C_t, K_{s_l} and K_{s_t}, whose values may be time variables. The variables g and a denote the gravitational and longitudinal accelerations of the vehicle body MC. The parameter $\delta_{(.)}$ denotes maximal possible displacement of the SS given by the corresponding elastic limiters [10]. The above relations can be simplified and presented in a vector-matrix form taking that the state coordinates are x_b, θ, x_{l_1} and x_{l_2}, and using the transformation expressions given in the form [10]:

$$x_{l_2} = x_b + l_1\theta$$
$$x_{t_2} = x_b - l_2\theta \tag{2.20}$$

Now the scalar dynamic equations of the system behavior (2.18) can be written as a vector, nonlinear, time-dependent second-order differential equation of the form [10]:

$$\mathcal{M}(t)\ddot{Z} + \mathcal{F}(Z, \dot{Z}, t) = \mathcal{G}U(t) + \mathcal{H}(t)W(t) + \mathcal{S}(t) \tag{2.21}$$

where the mentioned matrices and vectors are of the form:

$$\mathcal{M}(t) = \begin{bmatrix} M_s(t) & 0 & 0 & 0 \\ 0 & I(t) & 0 & 0 \\ 0 & 0 & M_{u_l}(t) & 0 \\ 0 & 0 & 0 & M_{u_t}(t) \end{bmatrix}; \quad Z = \begin{bmatrix} x_b \\ \theta \\ x_{l_1} \\ x_{t_1} \end{bmatrix};$$

$$
\mathcal{G}(t) = \begin{bmatrix} 1 & 1 \\ l_1 & -l_2 \\ -1 & 0 \\ 0 & -1 \end{bmatrix}; \quad \mathcal{H}(t) = \begin{bmatrix} 0 & 0 \\ 0 & 0 \\ K(t) & 0 \\ 0 & K(t) \end{bmatrix};
$$

$$
U(t) = \begin{bmatrix} u_l(t) \\ u_t(t) \end{bmatrix}; \quad W(t) = \begin{bmatrix} \omega(t) \\ \omega(t-\tau) \end{bmatrix}; \tag{2.22}
$$

The presented half-car vehicle model can be described in another, simpler way. If deflections of the SS and tire pneumatic are treated as a unique motion, i.e. if the equivalent deformations of the SS and tire are assumed, then the number of differential equations describing dynamic behavior of such a system is reduced by two. Such a model has been presented in [11, 12]. To use the simplified model it is necessary to identify the corresponding equivalent parameters of stiffness and damping which correspond to the characteristics of the unified system, SS + tire pneumatic.

In the plane perpendicular to the road surface there is also motion in the direction of changing the rolling angle of the vehicle body along its longitudinal axis. The previously described dynamic model (2.18) (i.e. (2.21)) can be relatively easily modified so that it describes the vehicle rolling dynamics [13].

2.2.3 Single Track Vehicle Model

Lateral vehicle dynamics and dynamics of motion about the vertical rotation axis are most often studied using the so-called "single-track vehicle model" [14, 15]. In the literature, this model is also known under the popular name "bicycle model of road vehicle" [16] or "two-wheel vehicle model" [17]. The alternative terms stem from the fact that the vehicle is considered as an object represented by a pair of wheels on a common longitudinal axis (Fig. 2.11). This model was developed by Reickert and Schunck [18], still in 1940. The model is especially convenient for studying the vehicle lateral dynamics.

Neglecting the dynamics of pitch, roll, and heave motion of vehicle body the vehicle model can also be defined in the road surface plane, i.e. in the plane of the vehicle motion. For this purpose it is possible to group the wheels on the front axle by assuming the existence of a virtual wheel placed at the front end of the longitudinal car axis, as presented in Fig. 2.11. The same can be done with the rear axle wheels. On adopting these approximations the vehicle body can be considered as being represented by a rigid beam [15] of mass m and inertia moment J about the vertical axis passing through the MC. Assuming that the longitudinal distribution of the vehicle mass is equivalent to the masses concentrated on the axes of front and rear wheels, the total vehicle mass m is related to the moment of inertia J as $J = m\, l_f\, l_r$ [14]. In the model definition use was made of the following notations: ψ is the yaw angle of the vehicle about its vertical axis passing through MC measured with respect to the fixed inertial coordinate frame $x_0 0 y_0$ attached to the ground. The angle ψ represents the angle between the axes x_0 and x of the fixed coordinate frame and the mobile coordinate frame $x0y$ attached to the MC; angle β is the side slide angle of the

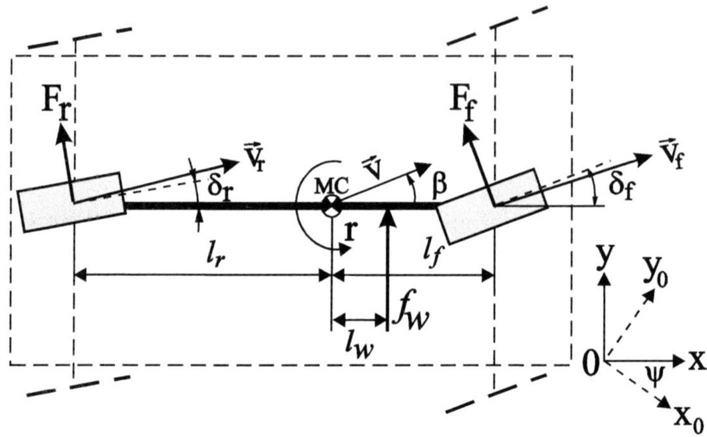

Figure 2.11: Single-track vehicle model with the possibility of changing ground steering angles of the front and rear tires

vehicle, i.e. the angle between the vehicle axis and the velocity vector $\vec{v}$; v is the velocity of MC motion; $r = \dot{\psi}$ is the angular velocity of vehicle yawing; a_f and a_r are respective lateral accelerations (along the Y-axis) at the front and rear vehicle axis; δ_f and δ_r are the respective ground steering angles of the front and rear wheels; l_r and l_f are the geometrical vehicle parameters representing the distance of MC from the rear and front wheel axle respectively. The span between the wheel axles represents the sum of the lengths $l = l_f + l_r$; f_w is the lateral force representing disturbance. It is assumed that this force is orthogonal to the longitudinal car axis; l_w is the distance of CM from the point at which the mentioned disturbance force f_w is acting.

Apart from the vector of external forces[10] f_w the system is also under action of the lateral forces F_f, F_r acting on the tires. Their direction is orthogonal onto the corresponding longitudinal tire axes (Fig. 2.11). In the first approximation of these forces by linear model[11], they can be calculated using the following relations:

$$
\begin{aligned}
F_f &= C_f\,\alpha_f \\
F_r &= C_r\,\alpha_r
\end{aligned}
\tag{2.23}
$$

where C_f and C_r are the so-called "cornering stiffness coefficients" of the tires.

---

[10]Components of this force are: the aerodynamic resistance force, force of side wind gust and the resistance force of the tire pneumatics rolling on the road surface.

[11]In the text to follow nonlinear tire model will be described in more details. For the current discussion it suffices to use a linear approximation of this force.

These coefficients vary to a certain extent in dependence of the value of slip friction coefficient μ between the road surface and tire pneumatics, so that it is possible to write: $F_f = F_f(\alpha_f, \mu)$ and $F_r = F_r(\alpha_r, \mu)$. The angles α_f and α_r are the characteristic tire slip angles. These angles delimit the space between the longitudinal axis of the tire and the direction of translational velocity (vectors $\vec{v}_f$ and $\vec{v}_r$ in Fig. 2.11) of the tire pneumatic measured at the tire center. Taking all this into account, the single-track vehicle model can be defined in the form of state equation and output equation of the system:

$$
\begin{aligned}
\dot{x} &= A\,x + B\,u + H\,f_w \\
y &= C\,x + D\,u + G\,f_w
\end{aligned}
\tag{2.24}
$$

Elements of the vectors and matrices in the model (2.24) are as follows [15]:

$$
x = \begin{bmatrix} \beta \\ r \end{bmatrix}; \quad y = \begin{bmatrix} \beta \\ a_f \end{bmatrix}; \quad u = \begin{bmatrix} \delta_f \\ \delta_r \end{bmatrix};
$$

$$
A = \begin{bmatrix} -\frac{C_f + C_r}{m\,v} & -\left(1 + \frac{C_f l_f - C_r l_r}{m\,v^2}\right) \\ -\frac{C_f l_f - C_r l_r}{J} & -\frac{C_f l_f^2 + C_r l_r^2}{J\,v} \end{bmatrix};
\tag{2.25}
$$

$$
B = \begin{bmatrix} \frac{C_f}{m\,v} & \frac{C_r}{m\,v} \\ \frac{C_f l_f}{J} & -\frac{C_r l_r}{J} \end{bmatrix}; \quad C = \begin{bmatrix} 1 & 0 \\ -\eta & -\frac{l_f}{v}\eta \end{bmatrix};
$$

$$
D = \begin{bmatrix} 0 & 0 \\ \eta & 0 \end{bmatrix}; \quad H = \begin{bmatrix} \frac{1}{m\,v} \\ \frac{l_w}{J} \end{bmatrix}; \quad G = \begin{bmatrix} 0 \\ \frac{l_r + l_w}{m\,l_r} \end{bmatrix}; \quad \eta = \frac{C_f\,(l_f + l_r)}{m\,l_r}
$$

The presented linear model is a good approximation of a real system if the following assumptions hold: (i) β, δ_f and δ_r have small amplitudes, (ii) the velocity value $v = const$ and, (iii) the dynamics of pitch, roll and heave negligibly affect lateral motion of the vehicle.

The characteristic polynomial of the linear model (2.24) can be written in the known form [14]:

$$
\begin{aligned}
p_0(s) &= \omega_0^2 + 2\,D_0\,\omega_0\,s + s^2 \\
\omega_0^2 &= \frac{C_r C_f l^2 + mv^2(C_r l_r - C_f l_f)}{m^2 v^2 l_r l_f} \\
D_0 &= \frac{l(C_r l_r + C_f l_f)}{2\sqrt{l_r l_f [C_r C_f l^2 + mv^2(C_r l_r - C_f l_f)]}}
\end{aligned}
\tag{2.26}
$$

where s is the Laplace operator, D_0 is the relative damping of the system yawing, and ω_0 is natural frequency of the system. The natural frequency decreases with increase in speed of the vehicle motion v. A satisfactory factor of relative yaw damping D_0 with fast cars can be attained by enhancing the span $l = l_r + l_f$ between the wheel axles [14].

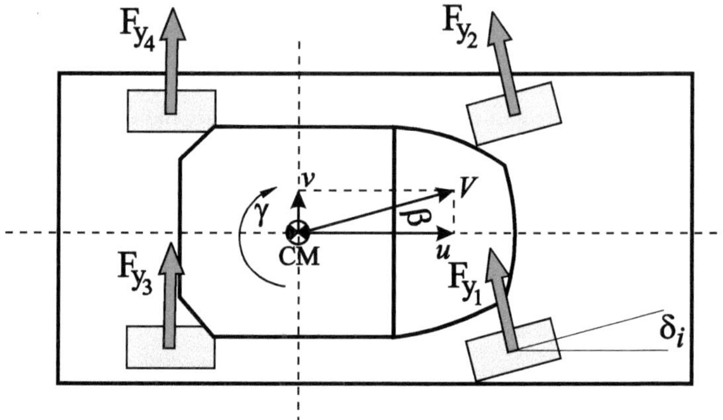

Figure 2.12: Planar vehicle model describing dynamics of vehicle motion in three coordinate directions: longitudinal, lateral, and yaw angle direction

2.2.4 Model of Longitudinal and Lateral Vehicle Dynamics

To the group of planar models belongs also the model of longitudinal and lateral vehicle dynamics. This model describes the vehicle dynamic behavior in the longitudinal and lateral directions as well as in the direction of the vehicle body yawing about the vertical axis passing through the vehicle body MC. In Fig. 2.12 is illustrated four-wheel planar model [19]. The dynamic model is described by three scalar differential equations of the behavior in each of the considered coordinate directions. These relations are of the following form [19]:

$$m\,(\dot{u} - v\,\gamma) \;=\; \sum_{i=1}^{4} F_{y_i}\,\sin(\delta_i)$$

$$m\,(\dot{v} + u\,\gamma) \;=\; \sum_{i=1}^{4} F_{y_i}\,\cos(\delta_i)$$

$$I_z\,\dot{\gamma} \;=\; \sum_{i=1}^{4} M_{z_i} \qquad (2.27)$$

In Fig. 2.12 and relations (2.27) use was made of the following notations: u is the longitudinal velocity of the vehicle; v is the lateral vehicle velocity defined with respect to the longitudinal vehicle axis; γ is the angular yawing velocity of the vehicle body about the vertical z-axis passing through the vehicle body MC; angle β represents the so-called vehicle side slip angle about the axis passing through the vehicle body MC. This angle is calculated from the relation $\beta = arctg(v/u)$; $\dot{\beta}$ is the corresponding rate of change of the side slip angle β; m and I_z are the respective mass and inertia moment about the yaw axis; V is the intensity of vehicle planar speed $V = \sqrt{u^2 + v^2}$; δ_i is the tire ground steering angle of the i-th tire for the possible value of the index $i = 1, 2$; F_{y_i} and M_{z_i} are the respective lateral yaw force on the tire and yaw moment produced by the i-th tire yaw force. In the sequel we will show that the tire yaw force F_{y_i} is a nonlinear function whose characteristics depend on the tire slipping ratio, tire side slip angle and vertical tire load force. The nonlinear tire characteristic gives a nonlinear character of the behavior of the considered vehicle model, presented in Fig. 2.12 and described by the relations (2.27).

2.3   Spatial Vehicle Models

The choice of the appropriate control strategy and the way of its realization represent a delicate problem, solution of which demands sufficiently deep knowledge of the dynamic behavior of the road vehicle under various motion conditions. Thus, a sufficiently reliable vehicle model is needed. The models used most often are simplified system models, such as for example the previously described quarter-car model [7], half-car model [10], planar model of lateral dynamics [19], model of longitudinal dynamics [20, 21], etc. A common characteristic of all the mentioned models is that none of them describes the overall dynamics of the vehicle, but only its partial dynamics.

2.3.1   Spatial 22 DOFs Vehicle Model

A vehicle as a real rigid-body system possesses over 30 DOFs of motion. If we consider the conventional spatial vehicle model shown in Fig. 2.13, we can distinguish two functional modules [22], [23]: the vehicle body (structure) and its suspension system. Thus, we can consider a vehicle body as a mechanical system representing a rigid body supported by its elastic suspensions [9]. Also we can consider the vehicle SS and the ground over which the vehicle moves as its "dynamic environment". This functional partitioning will be useful in the synthesis of dynamic control algorithms of the system motion. Such an approximation is justified by the fact that the vehicle moves over a rigid ground. Then, the dynamic characteristics of the vehicle SS influence more significantly the vehicle dynamic behavior in relation to the ground characteristics. In that case the road surface can be practically considered as a kinematic boundary for the vehicle during its motion. In this monograph, a relatively complex, nonlinear, tridimensional model of a road vehicle will be presented. The model that will be dealt with in this section is a good approximation of the considered real system. It describes sufficiently well the nonlinear character of the vehicle dynamics in a rather broad span of variation of state variables and input control/disturbance signals.

The complex nonlinear model used by Peng and Tomizuka in their simulation

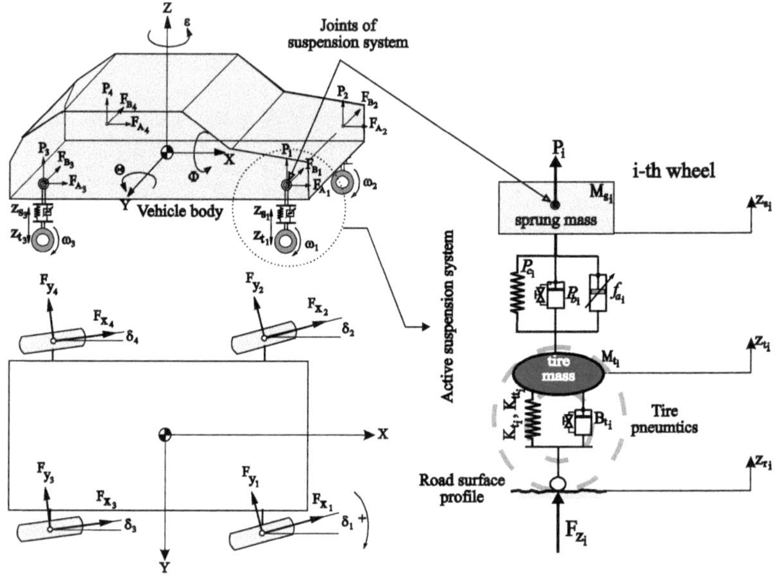

Figure 2.13: Functional scheme of the 3D-model of a road vehicle with 22 DOFs of motion

experiments [9], can be adopted, rearranged and extended, taking into account the dynamics of the suspensions and tires (see Fig. 2.13). This original model is based on the model developed by Lugner [22] and the tire model by Sakai [24]. An extended nonlinear model having 22 DOFs of motion, with the possibility of 4-wheel driving (4WD) and 4-wheel steering (4WS) will be described in this subsection. This model can be used for stability analysis, control synthesis, parameter estimation, and simulation [25], as will be described below. The model is illustrated in Fig. 2.13. The complex model describes the six DOFs of motion of the vehicle MC in three coordinate directions (x, y, z) and three rotations $(\phi, \theta, \varepsilon)$ of the vehicle about its main axes of inertia. The model also describes the dynamics of the 4-wheel SS in the vertical[12] direction, as well as the tire dynamics in the same direction. Each wheel, in addition to the vertical tire deflection possesses two extra DOFs: rotation about the horizontal axis at an angular velocity ω_i (for $i = 1, \ldots, 4$) and rotation δ_i (Fig. 2.13) about the vertical axis with respect to the road surface. The other rotation represents a

---

[12]We assume here the dynamics of the vehicle motion in the direction perpendicular to the horizontal surface of the road. If the road surface deviates for a small angle from the horizontal plane, the notion "vertical dynamics" should be taken as conditionally true.

change of the tire's ground steering angle.

2.3.1.1 Rigid Body Vehicle Dynamics

The vehicle body model is determined by its rigid body dynamics, and it can be expressed by the vector equation [25]:

$$H(q, d)\ddot{q} + h(q, \dot{q}, d) = \tau + F(q, \dot{q}, d) \tag{2.28}$$

where

$$\begin{aligned}
q &= [x \ y \ \varepsilon \ z \ \phi \ \theta]^T \\
\dot{q} &= [\dot{x} \ \dot{y} \ \dot{\varepsilon} \ \dot{z} \ \dot{\phi} \ \dot{\theta}]^T \\
\ddot{q} &= [\ddot{x} \ \ddot{y} \ \ddot{\varepsilon} \ \ddot{z} \ \ddot{\phi} \ \ddot{\theta}]^T
\end{aligned} \tag{2.29}$$

Here, q is a (6×1) vector of system state variables describing the position/orientation and velocity of the vehicle body MC with respect to the coordinate frame $0 \ X_0 \ Y_0 \ Z_0$ fixed to the ground (Fig. 2.14); x, y, z are the longitudinal, lateral, and vertical positions of the vehicle MC along the three coordinate directions, expressed in $[m]$; $\phi, \theta, \varepsilon$ are the corresponding angles of roll, pitch and yaw of the vehicle body in $[rad]$, and the relative angles measured between the fixed coordinate frame $0 \ X_0 \ Y_0 \ Z_0$ and the mobile coordinate frame $0 \ X \ Y \ Z$ placed at the vehicle body MC (Fig. 2.13). The fixed coordinate frame $OX_0Y_0Z_0$ (Fig. 2.14) is attached to a point of the road surface[13]. Orientation of the axes X_0, Y_0 of the fixed coordinate frame can be chosen in an arbitrary way. It is convenient to adopt that they coincide say with the geographic directions (directions of the parallels and meridians) as is presented in Fig. 2.14. The mobile coordinate system $OXYZ$ is attached to the vehicle body MC. This frame is movable and it moves together with the vehicle. Geometrical relations between these two coordinate frames determine relative change of the position and orientation of the vehicle in the space during motion; $H(q, d)$ is a (6×6) inertia matrix expressed in $[kg]$ and $[kgm^2]$ respectively; $h(q, \dot{q}, d)$ is a (6×1) vector of gravitational and centrifugal forces acting at the vehicle MC, expressed in $[N]$ and $[Nm]$ respectively; τ is a (6×1) vector of driving forces and torques referred to the vehicle MC, expressed in $[N]$ and $[Nm]$ respectively; $F(q, \dot{q}, d)$ is a (6×1) vector of the external forces and torques acting on the vehicle body during its motion along the road. Elements of this vector take into account forces and torques of tire rolling resistance, aerodynamic resistance forces during motion, and the damping torque of the yaw rate during cornering. The vector d represents an $l \times 1$ vector of the system parameters.

The above model holds in the case of adopting the following assumptions:

(i) The vehicle body from the point of view of mechanics represents a rigid body supported on four elastic subsystems that form the vehicle SS.

---

[13]Position of the fixed coordinate frame can be chosen arbitrarily in the space. It is commonly assumed that it coincides with the projection of the vehicle body MC on the road surface in the initial moment.

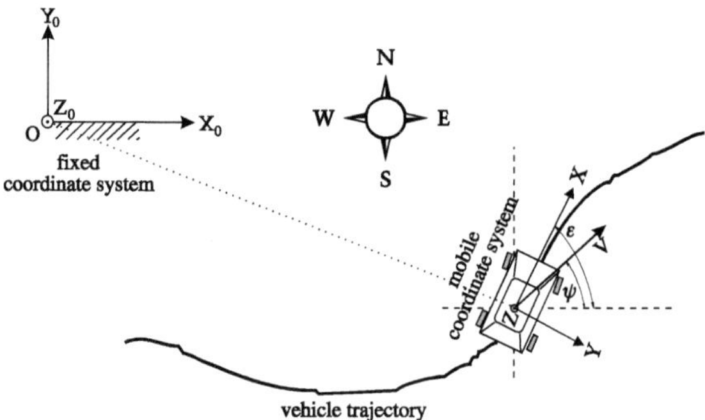

Figure 2.14: Coordinate frames used for defining the vehicle position and orientation during motion

(ii) The SS has elastic properties only in the "vertical direction" (perpendicular to the road surface), whereas it is considered absolutely rigid in the longitudinal and lateral directions.

(iii) Deviation of the relative position of the vehicle body during motion from its reference value when the vehicle is at rest with respect to the road plane is small. Concretely, in the case of small angles of pitch and roll the following approximations[14] $\phi \approx sin(\phi)$ and $cos(\phi) \approx 1$ hold.

(iv) The road surface is practically horizontal, i.e. side elevation angle of the road is small (not exceeding several degrees).

---

[14]Under the notion "small angle" we assume angles whose absolute values are less than 10 degrees, i.e. $0.1745\ [rad]$. The values of the trigonometric functions sine and cosine for this angle boundary value are: $\sin(0.1745) = 0.1736$ and $\cos(0.1745) = 0.9848$.

These assumptions are realistic for several reasons. Elastic properties of the vehicle body are relatively less pronounced than the elastic characteristics of SS and tire pneumatics. Elastic properties of SS, which is by virtue of its design in the longitudinal and lateral directions rigidly connected to the vehicle body, are negligible compared with the elastic characteristics in the vertical direction, where they are very pronounced. Roll and pitch angles of the vehicle body in the course of motion on a plain road are relatively small, amounting to not more than several degrees. Construction standards of contemporary highways prescribe small elevation angles of the motion surface with respect to the absolutely horizontal plane. Also, the contemporary roads are constructed so that the curve radii are not small, i.e. no "sharp" bends are allowed on such roadways.

In the model (2.28) use was made of the following notation:

- β: side slip angle $[rad]$ (see Fig. 2.15),

- ψ: orientation angle of the forward velocity vector $\psi = \varepsilon + \beta$ $[rad]$ (Fig. 2.15),

- γ: super-elevation angle of the road surface $[rad]$ (Fig. 2.15),

- $F_{A_i}(F_{B_i})$: longitudinal (lateral) force at the joint center of the i-th wheel SS $[N]$ (Fig. 2.13),

- P_i: vertical load force at the i-th joint of SS $[N]$ (Fig. 2.13),

- F_{z_i}: Vertical load force of the i-th tire $[N]$ (Fig. 2.13),

- M_x, M_y, M_z: moments about main axes of inertia (X, Y and Z axes) $[Nm]$,

- $l_1(l_2)$: distance of MC from the front (rear) vehicle axle $[m]$ (Fig. 2.15),

- h_1: distance of MC from the center of the i-th SS joint in the vertical direction $[m]$ (Fig. 2.15),

- h_2: distance of MC from the roll axis $[m]$ (Fig. 2.15),

- $h_4(h_5)$: $z(x)$-distance of MC from the pitch axis $[m]$ (Fig. 2.15),

- s_b: vehicle track $[m]$ (Fig. 2.15),

- m: lump mass of the road vehicle $[kg]$

- I_x, I_y, I_z: moments of inertia in the directions X, Y, Z $[kg\,m^2]$,

- K_x, K_y: coefficients of aerodynamic resistance to vehicle motion $[N/m^2]$,

- K_ε: yaw damping ratio of the vehicle about vertical axis $[Nm/(rad/s)]$,

- g: constant of gravitational acceleration $[m/s^2]$,

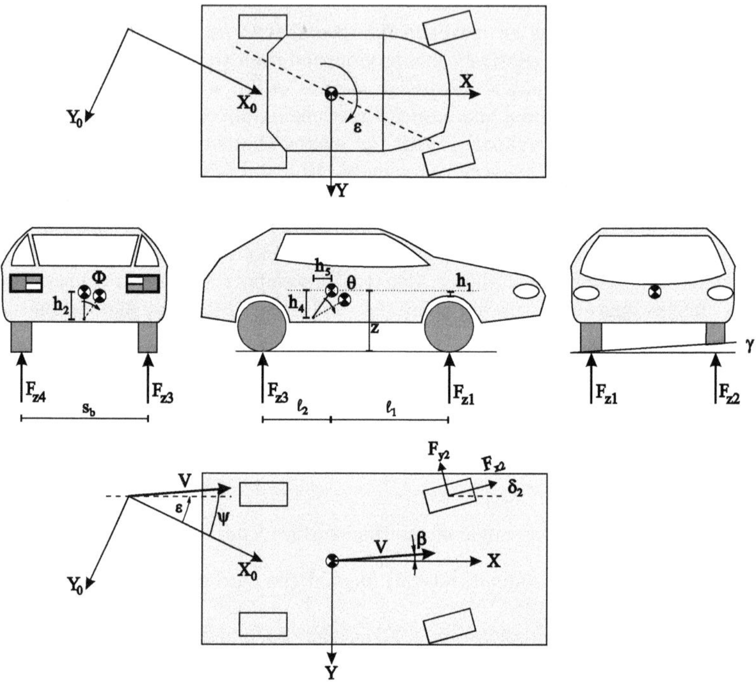

Figure 2.15: Geometrical vehicle parameters and characteristic angles used in the model

- V_i: translational velocity of the tire MC $[m/s]$,

- ζ_i: angle of orientation of the velocity vector of the i-th tire $[rad]$,

- f_{r_i}: tire rolling resistance, i.e coefficient of rolling friction.

Matrices and vectors in the model (2.28) are defined in the following way:

$$H(q,d) = \begin{bmatrix} m & 0 & 0 & 0 & 0 & m\,h_4 \\ 0 & m & 0 & 0 & -m\,h_2 & 0 \\ 0 & 0 & I_z & 0 & I_z\,\theta & -I_z\,\phi \\ 0 & 0 & 0 & m & 0 & -m\,h_5 \\ 0 & 0 & -I_x\,\theta & 0 & I_x & 0 \\ 0 & 0 & I_y\,\phi & 0 & 0 & I_y \end{bmatrix} \tag{2.30}$$

$$
h(q, \dot{q}, d) = \begin{bmatrix} -m\ddot{y}\dot{\varepsilon} + mg\gamma\beta \\ m\dot{x}\dot{\varepsilon} - mz\ddot{\gamma} + mg\gamma \\ I_z\theta\ddot{\gamma} - (I_x - I_y - I_z)(\dot{\gamma} + \dot{\phi})\dot{\theta} - I_z\dot{\phi}\dot{\theta} \\ -m\dot{x}(\dot{\gamma}\beta - \dot{\psi}\gamma) + m\dot{\gamma}\dot{y} + mg \\ I_x\ddot{\gamma} - (I_y - I_z + I_x)\dot{\theta}\dot{\varepsilon} \\ I_y\ddot{\gamma}\beta - I_y\gamma\ddot{\psi} + (I_x - I_y - I_z)\dot{\gamma}\dot{\varepsilon} + (I_x + I_y - I_z)\dot{\phi}\dot{\varepsilon} \end{bmatrix} \quad (2.31)
$$

$$
F(q, \dot{q}, d) = \begin{bmatrix} -K_x\dot{x}^2 - \sum_{i=1}^{4} f_{r_i}F_{z_i}\cos\zeta_i \\ -K_y\dot{y}^2 - \sum_{i=1}^{4} f_{r_i}F_{z_i}\sin\zeta_i \\ -K_\varepsilon\dot{\varepsilon} + M_{rr} \\ 0 \\ 0 \\ 0 \end{bmatrix} + F_w \quad (2.32)
$$

$$
M_{rr} = -l_1 \sum_{i=1,2} f_{r_i}F_{z_i}\sin\zeta_i + l_2 \sum_{i=3,4} f_{r_i}F_{z_i}\sin\zeta_i
$$
$$
+ \ 0.5\, s_b \sum_{i=1,3} f_{r_i}F_{z_i}\cos\zeta_i - 0.5\, s_b \sum_{i=2,4} f_{r_i}F_{z_i}\cos\zeta_i
$$

where M_{rr} is the moment of rolling resistance reduced at the vehicle MC, produced by yawing about the vertical Z-axis. F_w is the (6×1) vector of side force impact upon the vehicle body.

The system possesses three control variables on each wheel (see Fig. 2.13):
(i) ground steering angle δ_i,
(ii) tire angular velocity ω_i, and
(iii) active damping force f_{a_i} in the viscous cylinder of SS.
The values of subscript "i" change in the span $i = 1, \ldots, 4$.

The vector of generalized forces and torques τ in (2.28) can be calculated on the basis of the relation:

$$
\tau = [\sum_{i=1}^{4} F_{A_i} \quad \sum_{i=1}^{4} F_{B_i} \quad \mathcal{M}_Z + \theta\mathcal{M}_X - \phi\mathcal{M}_Y \quad \ldots
$$
$$
\ldots \quad \sum_{i=1}^{4} P_i \quad \mathcal{M}_X - \theta\mathcal{M}_Z \quad \mathcal{M}_Y + \phi\mathcal{M}_Z]^T \quad (2.33)
$$

where $\mathcal{M}_X$, $\mathcal{M}_Y$ and $\mathcal{M}_Z$ are the corresponding moments acting at the vehicle body MC about the axes of the coordinate frame $O\,X\,Y\,Z$ attached to the vehicle body MC (Fig. 2.13).

$$
\begin{aligned}
\mathcal{M}_X &= (\tfrac{s_b}{2} + h_2\phi)(P_1 + P_3) - (\tfrac{s_b}{2} - h_2\phi)(P_2 + P_4) + && (2.34) \\
&+ (h_1 - h_5\theta + l_1\theta)(F_{B1} + F_{B2}) + (h_1 - h_5\theta - l_2\theta)(F_{B3} + F_{B4}) \\
\mathcal{M}_Y &= (l_2 + h_4\theta)(P_3 + P_4) - (l_1 - h_4\theta)(P_1 + P_2) - && (2.35) \\
&- (h_1 - h_5\theta + l_1\theta)(F_{A1} + F_{A2}) - (h_1 - h_5\theta - l_2\theta)(F_{A3} + F_{A4}) \\
\mathcal{M}_Z &= (l_1 - h_4\theta)(F_{B1} + F_{B2}) - (l_2 + h_4\theta)(F_{B3} + F_{B4}) - && (2.36) \\
&- (\tfrac{s_b}{2} + h_2\phi)(F_{A1} + F_{A3}) + (\tfrac{s_b}{2} - h_2\phi)(F_{A2} + F_{A4})
\end{aligned}
$$

The generalized forces τ acting at the vehicle body MC are functions of the vector of control variables $\tau = \tau(\underline{\delta}, \underline{\omega}, \underline{f_a})$, where: vector of tire ground steering angles is $\underline{\delta} = [\delta_1 \; \delta_2 \; \delta_3 \; \delta_4]$, vector of the tire angular velocities $\underline{\omega} = [\omega_1 \; \omega_2 \; \omega_3 \; \omega_4]$, and vector of active damping forces in the cylinders $\underline{f_a} = [f_{a_1} \; f_{a_2} \; f_{a_3} \; f_{a_4}]$.

As a consequence of the action of the control variables $\underline{\omega}$, $\underline{\delta}$ and $\underline{f_a}$ the vehicle performs motion at a speed V which represents the so-called vehicle planar velocity in the plane defined by the coordinate system $O \; X_0 \; Y_0$ (Fig. 2.14).

$$\vec{V} = [\dot{x} \; \dot{y} \; 0]^T$$
$$V = \sqrt{\dot{x}^2 + \dot{y}^2} \tag{2.37}$$

The vehicle side slip angle (Fig. 2.15) is calculated from the difference of the angles:

$$\beta = \psi - \varepsilon \tag{2.38}$$

The velocity vector orientation angle of the vehicle (Fig. 2.14) is determined from the relation:

$$\psi = \arctan(\frac{\dot{y}}{\dot{x}}) \tag{2.39}$$

2.3.1.2 Vehicle Suspension Dynamics

The vehicle SS has a two-fold function: (i) damping of vertical oscillations of the vehicle caused by changing motion conditions and (ii) transmission of driving forces from the wheels onto the vehicle body. Forces at the SS joints act in three directions in regard to the longitudinal direction of motion X (Fig. 2.13). They are: vertical force P_i, longitudinal force F_{A_i}, and lateral force F_{B_i}. The ASS model is presented in Fig 2.13 and described by the relations (2.1)-(2.5). These relations describe a nonlinear system behavior. The sprung mass M_{s_i} of the i-th tire SS represents a fictitious system parameter. It can be explained in the following way. Let observe the vehicle body as a spatial beam with mass m. It is supported on the ground at four points. Then the sprung masses M_{s_i} of the particular tires are determined from the relations:

$$M_{s_i} = \frac{l_2}{2(l_1 + l_2)} m \quad \text{for} \quad i = 1, 2$$
$$M_{s_i} = \frac{l_1}{2(l_1 + l_2)} m \quad \text{for} \quad i = 3, 4 \tag{2.40}$$

where l_1 and l_2 are the spans between the points of support and vehicle MC, i.e. the distances between the vehicle MC and the axle of the pairs of front and rear tires.

The kinematic relations that concern the geometry of functional relations between the vehicle body MC position and vertical displacements at particular

SS joints are:

$$z_{s_i} = z - z_0 - h_5\theta - l_1\theta \pm \frac{s_b}{2}\phi \quad \text{for} \quad i = 1,2$$

$$z_{s_i} = z - z_0 - h_5\theta + l_2\theta \pm \frac{s_b}{2}\phi \quad \text{for} \quad i = 3,4 \qquad (2.41)$$

where: z_{s_i} and z_{t_i} are the respective relative positions of the joint center of the i-th tire SS with respect to the road surface and position of the i-th tire MC with respect to the same reference plane (see Fig. 2.13); z_0 is the relative MC height of the vehicle body when it is in the state of rest; z is the current relative height of the same point (Fig. 2.15). The quantities h_5 i s_b represent geometrical parameters of the vehicle, namely: the distance of the vehicle MC from the pitch axis in the X-direction and vehicle track.

The payload P_i of the i-th joint of SS (Fig. 2.13) and tire load F_{z_i} can be calculated on the basis of the following relations:

$$P_i = P_{c_i} + P_{b_i} - f_{a_i} \quad \text{for} \quad i = 1,\ldots,4$$

$$F_{z_i} = M_t\ddot{z}_{t_i} + T_{b_i} + T_{c_i} \quad \text{for} \quad i = 1,\ldots,4 \qquad (2.42)$$

The longitudinal F_{A_i} and lateral F_{B_i} forces act at the i-th joint of SS (Fig. 2.13) and they are transmitted to the vehicle body, to change its direction of motion and orientation in the space. These forces are nonlinear functions of global state coordinates q, their derivatives $\dot{q}$, $\ddot{q}$, control variables u_i, geometrical and dynamic system parameters d and the parameters of tire-road interaction $\tilde{d}$. This can be mathematically expressed in a general form as:

$$F_{A_i} = f(q,\dot{q},\ddot{q},u_i,d,\tilde{d}) \quad i = 1,\ldots,4$$

$$F_{B_i} = f(q,\dot{q},\ddot{q},u_i,d,\tilde{d}) \quad i = 1,\ldots,4 \qquad (2.43)$$

where the control variables u_i are the following quantities:

- δ_i - ground steering angle of the i-th tire,

- ω_i - angular velocity of the i-th tire rotation,

- f_{a_i} - active damping force in the viscous cylinder of the i-th SS.

Vector of control variables can be compactly presented in the following form:

$$\underline{u} = [u_1 \ u_2 \ u_3 \ u_4]^T$$

$$u_i = [\delta_i \ \omega_i \ f_{a_i}]^T \quad i = 1,\ldots,4 \qquad (2.44)$$

Tho block-scheme of a tridemensional vehicle model is presented in Fig. 2.16.

The longitudinal and lateral forces F_{A_i} and F_{B_i} at the SS joints (Fig. 2.13) can be calculated on the basis of the expressions [9]:

$$F_{A_i} = F_{x_i}\cos(\delta_i) - F_{yi}\sin(\delta_i) \quad i = 1,\ldots,4$$

$$F_{B_i} = F_{x_i}\sin(\delta_i) + F_{yi}\cos(\delta_i) \quad i = 1,\ldots,4 \qquad (2.45)$$

where F_{x_i} and F_{yi} are the longitudinal and lateral forces of the tires (Figs. 2.13 and 2.15), i.e. the forces of interaction between the i-th tire and road surface.

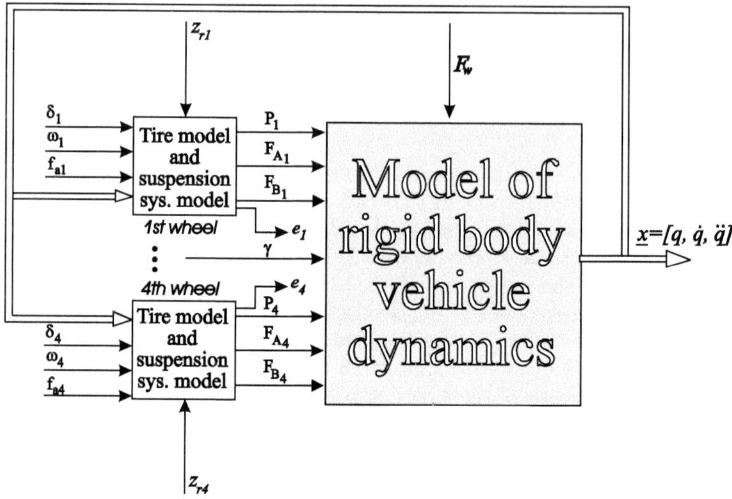

Figure 2.16: Block-scheme of spatial model of the road vehicle with 22 DOFs of motion suitable for open-loop simulation

2.3.1.3 Dynamics of Tire-Road Interaction

Tire characteristics have a dominant effect on nonlinear behavior of the road vehicle. For this reason it is especially important to know the properties of tire pneumatics because they are the sites of action of the forces and moments arising from the different conditions of motion on the road surface. Hence, great attention is paid to studying and modeling of the behavior of these vital elements of the system. However, we are not going to describe in detail the construction of tire pneumatics. Our attention will be focused on the problems of modeling tire forces and moments. The tire forces and moments depend on the tire construction, on the tread surfaces of pneumatics and of parameters of the tire-road interaction. There are several models describing tire forces and moments and they have been developed with the aim of studying the influence of tires on the dynamics of vehicle motion. Some of these models are based on the description of the physical nature of tire pneumatics [24, 26]. Other models are basically empiric. Almost all models whose aim is to provide a more or less exact description of the measured characteristics, rely in part upon empirical formulations [27]. In the contemporary literature one can find several tire models; the three most often encountered being:

    (i) Sakai's nonlinear tire model,

    (ii) "tire brush model", and

(iii) "magic formula" tire model.

Sakai's tire model: Dynamic behavior of tire pneumatics under different motion conditions (dry, slippery road) is difficult to describe in an exact way by mathematical relations. The existing models are mainly based on empirical patterns derived on the basis of experimental measurements. One of tire models that is frequently used in the analysis of dynamic behavior of road vehicles is the Sakai's nonlinear model [24, 26]. This model is a nonlinear transcendental function of several dependent variables [9]. This gives to the model a pronounced complexity of its form.

During the vehicle motion on the road there are the following tire interaction forces: tire load F_{z_i} (see Fig. 2.15), longitudinal tire force F_{x_i} acting in the direction of longitudinal axis of the tire pneumatic (Fig. 2.13) and the corresponding lateral interaction force F_{y_i} which is perpendicular to the direction of tire translation. The force of "vertical" tire load is calculated using the relation (2.6). The longitudinal F_{x_i} and lateral F_{y_i} tire forces are nonlinear functions of the variables: tire slip ratio s_i and the corresponding tire slip angle α_i [9]. The existing tire forces are also dependent of the pneumatic parameters and parameters of tire-road interaction. For the case of braking and traction these forces are determined from the following relations [9]:

During braking

$$
\begin{aligned}
F_{x_i} &= k_{t_i} s_i (1 - q_{t_i})^2 \cos(\alpha_i) + F_{z_i} \nu_{d_i} q_{t_i}^2 (3 - 2q_{t_i}) h_{t_i} s_i \\
F_{yi} &= k_{t_i} (1 - s_i)(1 - q_{t_i})^2 \sin(\alpha_i) + \\
&\quad + F_{z_i} \nu_{d_i} q_{t_i}^2 h_{t_i} (3 - 2q_{t_i}) \tan(\alpha_i)
\end{aligned}
\tag{2.46}
$$

During traction

$$
\begin{aligned}
F_{x_i} &= k_{t_i} s_i (1 - q_{t_i})^2 + F_{z_i} \nu_{d_i} q_{t_i}^2 (3 - 2q_{t_i}) h_{t_i} s_i \\
F_{yi} &= k_{t_i} (1 - s_i)(1 + s_i)(1 - q_{t_i})^2 \tan(\alpha_i) + \\
&\quad + F_{z_i} \nu_{d_i} q_{t_i}^2 h_{t_i} (3 - 2q_{t_i}) \tan(\alpha_i)
\end{aligned}
\tag{2.47}
$$

The notations used are: ν_{d_i} is the dynamic coefficient of tire slip friction; k_{t_i}, h_{t_i} and q_{t_i} are the intermediary parameters of the tire model depending on the degree of tire slip ratio, tire side slip angle, and on the parameters of tire-road surface interaction; effect of resistance of the tire pneumatic is included in the vector of external resistance forces $F(q, \dot{q}, \tilde{d})$ (2.32) and, as has been shown, it depends on the coefficient of tire rolling resistance f_{r_i}. The tire forces (F_{x_i}, F_{y_i}, F_{z_i}) are by means of SS transmitted onto the vehicle body, changing thus its motion direction and acceleration.

Side slip angle of the i-th tire is determined from the relation:

$$
\alpha_i = \delta_i - \zeta_i
\tag{2.48}
$$

Velocity orientation angle of the i-th tire is:

$$
\tan(\zeta_1) = \frac{\dot{y} + l_1 \dot{\varepsilon}}{\dot{x} - \frac{s_b \dot{\varepsilon}}{2}} \qquad \tan(\zeta_2) = \frac{\dot{y} + l_1 \dot{\varepsilon}}{\dot{x} + \frac{s_b \dot{\varepsilon}}{2}}
$$

$$\tan(\zeta_3) \;=\; \frac{\dot{y} - l_2\dot{\varepsilon}}{\dot{x} - \frac{s_b\dot{\varepsilon}}{2}} \qquad \tan(\zeta_4) = \frac{\dot{y} - l_2\dot{\varepsilon}}{\dot{x} + \frac{s_b\dot{\varepsilon}}{2}} \tag{2.49}$$

Tire slip ratio is defined as:

$$s_i \;=\; \frac{V_i\cos(\alpha_i) - r_i\omega_i}{V_i\cos(\alpha_i)} \qquad \text{during braking}$$

$$s_i \;=\; \frac{V_i\cos(\alpha_i) - r_i\omega_i}{r_i\omega_i} \qquad \text{during traction} \tag{2.50}$$

where V_i is the translational velocity of the i-th tire MC. It is determined from one of the available relations in dependence of tire disposition[15]:

$$V_1 \;=\; [(\dot{y} + l_1\dot{\varepsilon})^2 + (\dot{x} - \frac{s_b\dot{\varepsilon}}{2})^2]^{0.5}$$

$$V_2 \;=\; [(\dot{y} + l_1\dot{\varepsilon})^2 + (\dot{x} + \frac{s_b\dot{\varepsilon}}{2})^2]^{0.5}$$

$$V_3 \;=\; [(\dot{y} - l_2\dot{\varepsilon})^2 + (\dot{x} - \frac{s_b\dot{\varepsilon}}{2})^2]^{0.5}$$

$$V_4 \;=\; [(\dot{y} - l_2\dot{\varepsilon})^2 + (\dot{x} + \frac{s_b\dot{\varepsilon}}{2})^2]^{0.5} \tag{2.51}$$

The symbols r_i and ω_i represent the effective rolling radius of the i-th tire pneumatic and the angular velocity of it.

Effective tire rolling angle is calculated from the relation:

$$r_i = r_{i_0} - e_{t_i} \quad i = 1,\ldots,4 \tag{2.52}$$

where r_{i_0} is the original radius of the i-th tire and e_{t_i} is the tire pneumatic deflection $e_{t_i} = z_{t_i} - z_{r_i}$ (Fig. 2.2), dependent of the vertical load of the tire pneumatic F_{z_i}.

Tire contact length is determined from the relation:

$$l_{t_i} = 2(r_{i_0}^2 - r_i^2)^{0.5} \quad i = 1,\ldots,4 \tag{2.53}$$

The above-mentioned intermediary parameters of the model are defined with the aim of simplifying the final expressions for the longitudinal F_{x_i} and lateral F_{y_i} tire forces in the relations (2.46) and (2.47):

$$k_{t_i} \;=\; \frac{C_i\, w_{t_i}\, l_{t_i}^2}{2}$$

$$h_{t_i} \;=\; [(\tan(\alpha_i))^2 + s_i^2]^{-0.5}$$

$$q_{t_i} \;=\; \frac{k_{t_i}}{3\, \nu_{s_i}\, F_{z_i}\, h_{t_i}} \tag{2.54}$$

The parameters determining the tire-road surface interaction are:

- RC_i: characteristic coefficient of tire-road interaction,

---

[15]Formal numbering of tires is represented in Figs 2.13 and 2.15

- $C_i = C_{i0} \cdot RC_i$: stiffness constant characterizing contact of the tire pneumatic and road surface,

- $\nu_{s_i} = \nu_{s_0} \cdot RC_i$: static friction coefficient,

- $\nu_{d_i} = \nu_{d_0} \cdot RC_i$: dynamic sliding friction coefficient,

- l_{t_i}: contact length of the i-th tire with ground,

- w_{t_i}: width of the i-th tire.

The presented nonlinear tire model is convenient for the analysis of the influence of particular parameters (e.g. coefficient of tire-road sliding friction) and characteristic variables (slip ratio, side slip angle, tire load) on the amplitudes of tire forces. In order to get a better insight into the nonlinear characteristics of tire pneumatics we shall present here several characteristic simulation results. Tire parameters used in model simulations are given in Appendix at the end of the chapter. In Fig. 2.17 are illustrated some of the results obtained by the model simulation. In Fig. 2.17a are presented functional relationships of the forces F_{x_i} and F_{y_i} concerning change of the tire slip angle α_i in the case of different values of sliding friction coefficient $\nu_{d_i} = \{0.3, 0.4, 0.5, 0.6 \text{ and } 0.7\}$. The shapes of these curves suggest a nonlinear character of the considered model. As far as the force intensities are concerned, it is evident from the graph that they are greater for the greater values of friction coefficients, and vice versa. Similar conclusion can also be drawn about the functional relationship force-tire slip ratio (Fig. 2.17b). The curves of the graphs $F_{y_i} - \alpha_i$, $F_{x_i} - s_i$ and $F_{y_i} - s_i$ exhibit a saturation trend at higher values of the independent variables, namely at the friction coefficient value $\nu_{d_i} > 0.5$. The graph in Fig. 2.17c illustrates the dependence of the force amplitude on the tire load: the larger is the tire load the greater are the values of longitudinal and lateral forces. The conclusion concerning the influence of the resistance parameter ν_{d_i} on the intensities of the forces F_{x_i} and F_{y_i} at the different tire load force F_{z_i} can be expressed in the following way - to the higher values of friction coefficient correspond larger longitudinal and lateral tire forces (Fig. 2.17c).

Tire brush model: It is known that the occurrence of lateral tire forces is related to the existence of the so-called selfaligning torque about the tire vertical axis [28]. This torque arises due to the fact that the lateral force[16] F_y acts at a point displaced for the distance t from the pneumatic center C (see Fig. 2.18) [28, 29]. This moment lever arm is called the tire pneumatic trail and it represents an important characteristic of tire pneumatic.

Using the SAE convention on axes, the selfaligning torque is determined from the relation [28]:

$$M_\delta = t \cdot F_y \tag{2.55}$$

---

[16]More exactly, this is the resultant of the lateral forces formed by integration of the local values of force amplitudes over the tire-ground contact range. Load distribution in this region resembles brush hairs, hence the model's name.

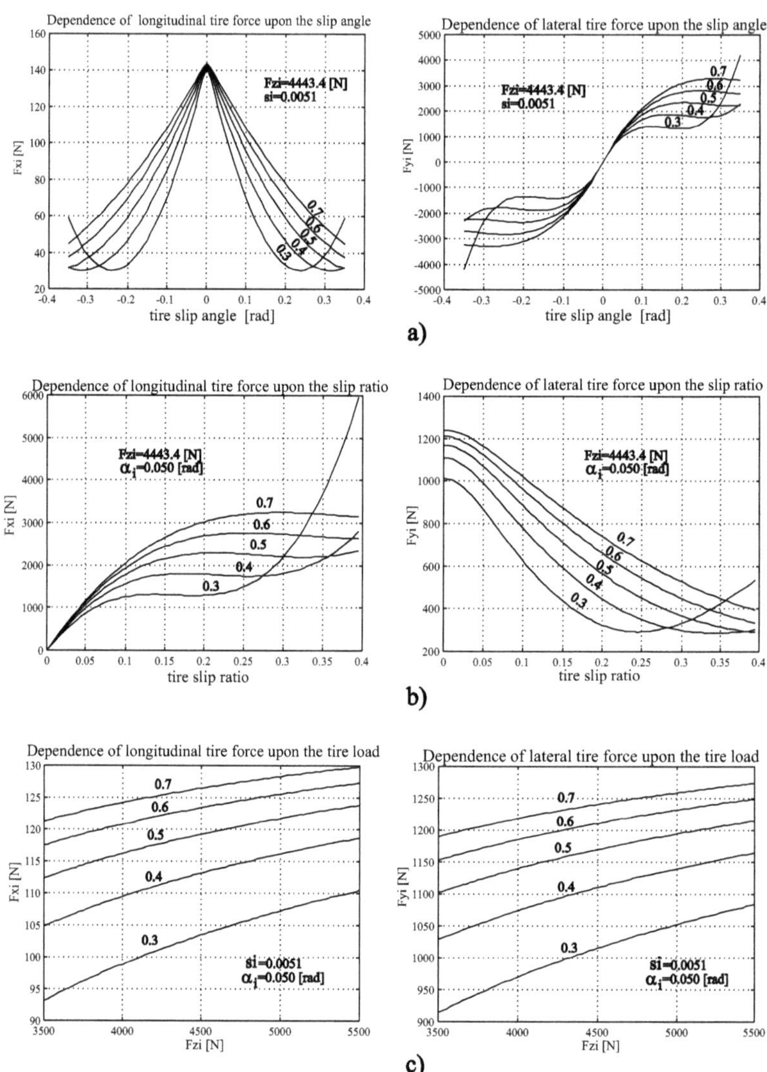

Figure 2.17: Functional dependence of tire forces on: a) tire side slip angle, b) tire slip ratio, and c) tire load for cases of different values of slipping friction coefficient

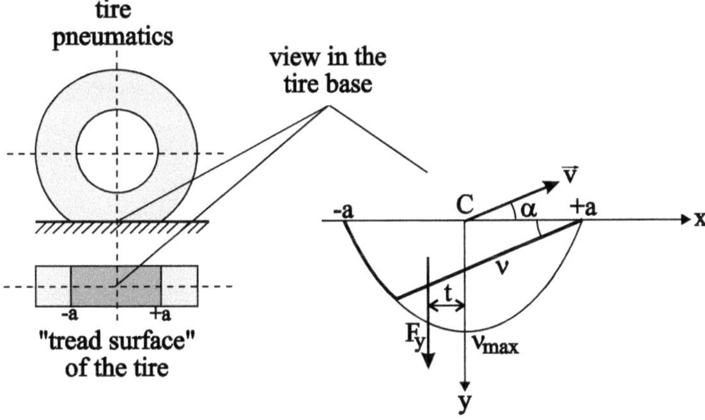

Figure 2.18: View in the tire base - tire load distribution at the vehicle yawing

In the case when the lateral force F_y and the moment M_δ are positive the tire pneumatic trail t is negative (Fig. 2.18). Let us shed some light on the tire pneumatics trail using the tire brush model. In Fig. 2.18 is shown the contact line of the tire model in stationary side slip. It is assumed that the tire load distribution q_z along the contact line is parabolic. The contact line is a straight line parallel to the velocity vector $\vec{v}$ where there exists adhesion between the tire tread elements and road surface (Fig. 2.18). The contact line is a curved line in the region of sliding occurrence [28]. In the adhesion region, the local side force changes linearly in dependence of the deformation ν of tread elements until a limiting value of adhesion force is attained. At the rear part of the tread surface the pneumatic trail elements slip on the road surface. The local side force is then determined by the sliding friction coefficient and the surface distribution of the load. The load distribution q_z, deformation of tread elements ν, and the tire local side force F_y^{elem} are determined from the relations:

$$
\begin{aligned}
q_z &= \frac{3\,F_z}{4\,u}\left\{1-(\frac{x}{u})^2\right\} \\
\nu &= (a-x)\tan(\alpha) \\
F_y^{elem} &= \begin{cases} c_y^{elem}\cdot\nu & \text{for } \; c_y^{elem}\cdot\nu \le \mu q_z \quad \text{adhesion} \\ \mu q_z & \text{for } \; c_y^{elem}\cdot\nu > \mu q_z \quad \text{sliding} \end{cases}
\end{aligned}
\qquad (2.56)
$$

where: c_y^{elem} is the stiffness of the pneumatic micro-particles, μ is the sliding friction coefficient, and α is the tire side slip angle [28]. In modeling, one should determine the equivalent resultant lateral tire force which is distributed along the contact line, equilibrating the forces and moments about the vertical z-axis of the pneumatic. Because of the asymmetry of distribution of the local side forces, the equivalent side force acts at a point displaced from the center "C" of the contact line for the distance t. In this way the force F_y produces the selfaligning torque M_δ. Finally, the considered tire model can be described by the relations:

$$F_y = \int_{-a}^{+a} F_y^{elem}\, dx$$

$$M_\delta = \int_{-a}^{+a} F_y^{elem} x\, dx$$

$$t = \frac{M_\delta}{F_y} \tag{2.57}$$

The change of the friction coefficient μ from the value μ_1 to the value μ_2 influences not only the change of the amplitude of the equivalent side tire force but also the value of the pneumatic trail t. In Fig. 2.19 is illustrated the effect of change of the friction coefficient μ on the mentioned parameters. In

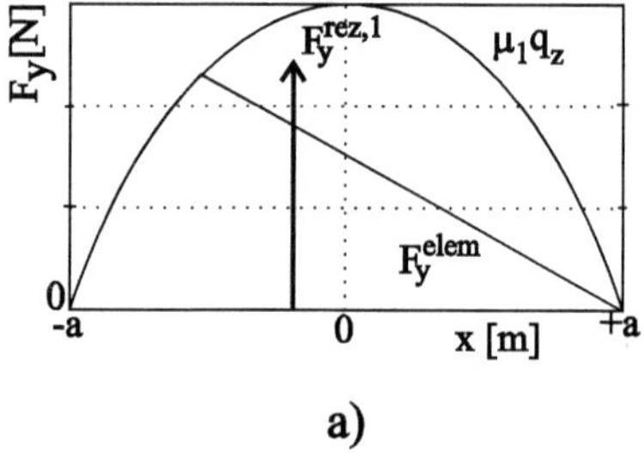

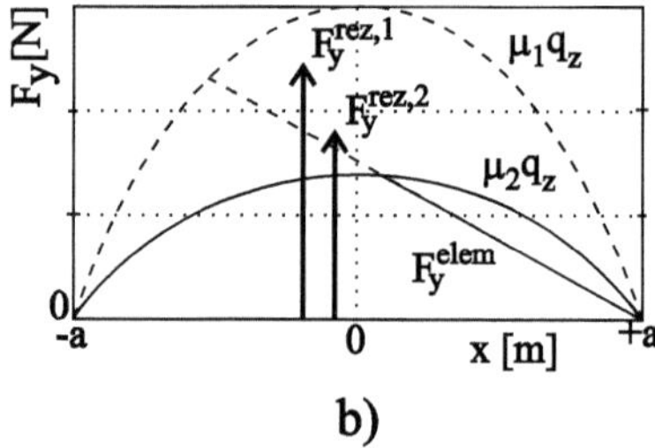

Figure 2.19: Effect of change of friction coefficient on the amplitude of lateral tire forces and the value of pneumatic trail

Fig. 2.19a is shown the distribution of the side tire force F_y for the values of

friction coefficient $\mu_1 > \mu_2$, whereas in Fig. 2.19b are given (for comparison) the changes of amplitudes of lateral forces along the tire contact line and the change of pneumatic trail value as a function of the change of the friction parameter $\mu = \mu_1$ and $\mu = \mu_2$.

In drawing the diagrams of the dependence of the lateral tire force F_y and selfaligning moment M_δ on the pneumatic side slip angle α for the different values of the sliding friction coefficient μ (Fig. 2.20a), there appeared the need of constructing a more complex diagram, the so-called Gough's diagram [28]. This diagram represents a useful means which can be used to estimate parameters of the tire-ground interaction, described in detail in [28, 30]. In Fig. 2.20b is given an example of Gough's diagram obtained by simulation of tire brush model using the following values of pneumatic parameters: $c_y^{elem} = 1.95705 \cdot 10^6$ $[N/m]$, $a = 0.086\ [m]$ and the value of vertical tire load $F_z = 4443.4\ [N]$. In the diagram, there are two parametric curves corresponding to constant values of tire pneumatic trails t and the sliding friction coefficient μ of the pneumatic on the ground. This diagram can serve for determining values of the unknown variables α and μ. This is feasible provided we know the values of amplitudes of the force F_y and moment M_δ, which serve as the basis for determining the value of tire pneumatic trail t. Total adhesion of the tire pneumatic exists in the region where the lines of constant friction coefficients μ almost coincide. From a theoretical point of view, total adhesion exists only when the slip angle vanishes, i.e. $\alpha \to 0$, or in the case of a theoretically infinite value of the friction coefficient μ. Along the ordinate axis shown in the diagram (Fig. 2.20b) the lines of constant slip angles coincide. This case is defined as the case of total tire slip. Taking into account what has been said above, the following conclusion can be drawn. Having at the disposal a relatively exact tire model and measuring the real values of tire forces and moments it is possible to estimate the unknown values of the variables α and μ. The knowing of these values is indispensable in calculating control of the lateral motion of the vehicle.

The tire brush model is more of a theoretical than of practical importance. It is aimed at a better understanding of the physical micro-phenomena arising in the contact of tire pneumatics and road surface. Real behavior of tire pneumatics is still rather different (because of the approximations adopted in the model) from the results obtained by simulation of the tire brush model. The main reasons for the existence of discrepancy is in the non-constancy of friction coefficients, lateral deformability of the body and outer belt of the pneumatic, and variation in pneumatic pressure distribution [30]. There are also some other complicating factors such as tire load variation, variations of the motion speed, tire inclination angle, combined longitudinal and lateral slipping, residual forces and moments, and the dynamic behavior of pneumatics [30]. For these reasons, this model has a more academic than practical importance in the research. It represents a useful tool enabling an easier understanding of the phenomena arising in tire pneumatics during the vehicle motion. To simulate models with a higher degree of exactness it is necessary to use tire models that are closer to a real behavior of the object. For this reason the Sakai model and the so-called "magic formula" model are more used in practice, as they are based on

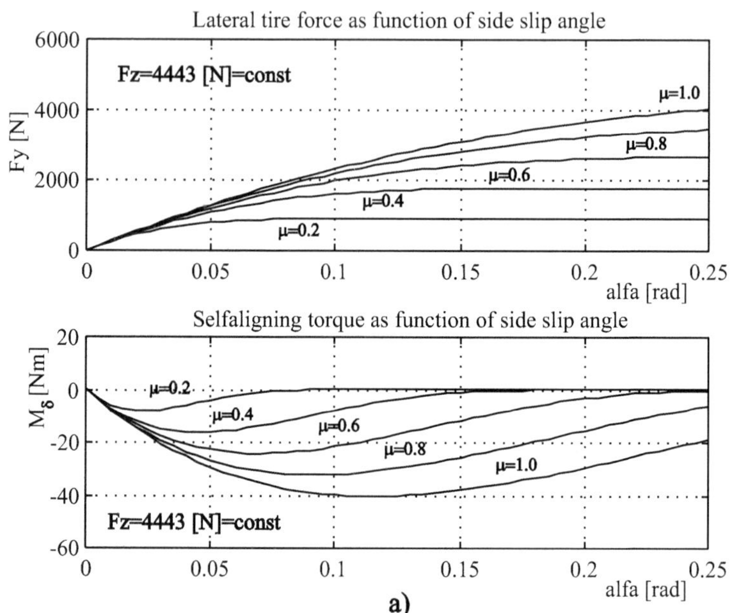

Lateral tire force as function of side slip angle
Fz=4443 [N]=const
μ=1.0
μ=0.8
μ=0.6
μ=0.4
μ=0.2
Fy [N]
alfa [rad]
Selfaligning torque as function of side slip angle
μ=0.2
μ=0.4
μ=0.6
μ=0.8
μ=1.0
Fz=4443 [N]=const
M_δ [Nm]
alfa [rad]
a)

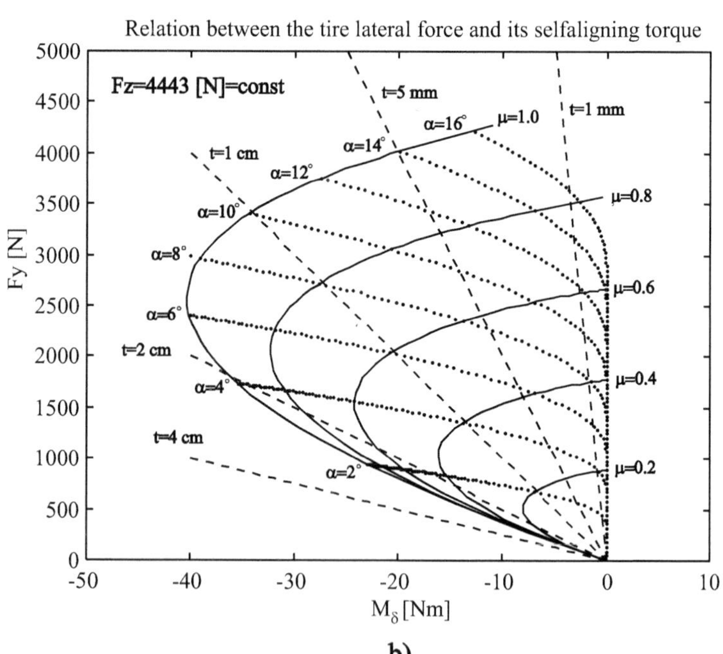

Relation between the tire lateral force and its selfaligning torque
Fz=4443 [N]=const
t=5 mm
t=1 mm
t=1 cm
α=16°
α=14°
α=12°
μ=1.0
α=10°
α=8°
μ=0.8
α=6°
t=2 cm
μ=0.6
α=4°
μ=0.4
t=4 cm
α=2°
μ=0.2
Fy [N]
M_δ [Nm]
b)

experimental measurement results.

The "Magic Formula" Tire Model: This model has been used most often in research during the last ten years. It is partly based on the apprehension of physical properties of tire pneumatics (like the Sakai's model), but more importantly, on empirical results. Pacejka and Bakker [27, 31] developed this model in co-operation with Volvo Car Corporation. The model enables description of the characteristic tire forces and moments in a steady-state regime of motion with a relatively high accuracy. It is used to study vehicle dynamics and in simulations. The model represents an empirical expression composed so to approximate results obtained by experimental measurements on real systems, with the aim of attaining a highest possible degree of congruence [28]. The kernel of the model is a mathematical formula known in the literature as "magic formula". This formula enables the calculation of tire forces and moments during motion involving pure slipping of pneumatics on the ground, both during cornering and braking/accelerating [27]. With the aim of attaining an exact description of the actual tire forces and moments under combined conditions of slipping (let say, in case of simultaneous braking and cornering), the model possesses the corresponding extensions of the basic formula. The purpose of these extensions is to enable combining of the basic formulas defined for the conditions of occurrence of pure sliding, for the sole braking, and tire cornering only.

The general expression of the "magic formula" describing either the sole lateral or longitudinal sliding can be represented by the following mathematical relation [28, 27]:

$$y(x) = D \sin \left[C \arctan \left\{ Bx - E(Bx - \arctan(Bx)) \right\} \right] \tag{2.58}$$

where:

$$\begin{aligned} Y(X) &= y(x) + S_v \\ x &= X + S_h \end{aligned} \tag{2.59}$$

The curves obtained for certain concrete values of the coefficients B, C, D and E are presented in Fig. 2.21a. The "magic formula" model produces curves passing through the origin of the coordinate frame $x = y = 0$, attain a maximum, and then approach a horizontal asymptote. For constant values of the coefficients B, C, D and E, the defined dependence curve represents an asymmetric function with respect to the frame origin, shifted in the vertical and horizontal directions by the respective offset values S_v and S_h. In order to allow the curve "express" its change of the offset value with respect to the origin an "offset" coordinate frame $Y(X)$ is introduced, which is shifted with respect to the frame $y(x)$.

The formula presented by the expressions (2.58) and (2.59) enables one to generate tire characteristics that with a high degree of congruence approximate the experimental curves: the lateral tire force F_y, the selfaligning moment M_δ and the longitudinal tire force F_x. These variables are functions of the side slip angle and tire slip ratio, described previously by the relations (2.48) and (2.50).

The output variable $Y(X)$ of the model (2.59) represents one of the variables F_y, F_x or M_δ. As argument X one can take the values of the variables α or s.

The offsets S_v and S_h appear as a result of the involvement of the effects appearing as a consequence of the existence of inclination arising from an improper mounting of the tire on the axle (Fig. 2.21b) for the amount of a small angle γ_c. The angle of tire inclination is equal to the sum of the following angles:

$$\gamma_c = \gamma_b + \phi \tag{2.60}$$

where: γ_c is the tire inclination angle with respect to the road surface; γ_b is the angle of tire inclination with respect to the vehicle body and ϕ is the vehicle body rolling angle. The existence of the angle γ_c causes shifts of the curves with respect to the coordinate frame origin (diagram in Fig. 2.21a). Similarly, the tire rolling resistance introduces shifts in the graph of the longitudinal force characteristics.

In the diagram presented in Fig. 2.21a, the coefficient D represents a maximal, peak value of the function $y(x)$ along the x-axis, and the product $(B \times C \times D)$ corresponds to the curve slope at the coordinate frame origin $x = y = 0$. This product represents the important coefficients: the so-called cornering stiffness $C_{F\alpha}$ and selfaligning stiffness $C_{M\alpha}$. They are important for linear approximations of the lateral force ($F_y = C_{F\alpha}\,\alpha$) and selfaligning torque ($M_\delta = C_{M\alpha}\,\alpha$). The factor C in (2.58) determines the boundary of the sine function in the formula, and hence it determines the shape of the resultant curve. Typical numerical values of this factor are: 1.30 for the lateral force F_y, 2.40 for selfaligning torque M_δ, and 1.65 for the characteristic of the braking force F_x. The factors D and C are also called "peak value factor" and "shape factor", respectively. The role of the factor B has to "control" the curve slope at the coordinate frame origin $x = y = 0$. For this reason it is called "stiffness factor". The remaining factor E, which is called "inflexion factor", is essential because it defines the shape of the curve in the vicinity of the peak value. At the same time, the factor E controls the sliding value x_m at which appears the "peak value" $D = y_m$, if this exist. The value of the coefficient E is determined from the following relation [27]:

$$E = \frac{Bx_m - \tan(\frac{\pi}{2C})}{Bx_m - \arctan(Bx_m)} \tag{2.61}$$

and it must not exceed "1" in order the curve could attain its maximal value $y = D$ [27]. The asymptotic value the function y converges at higher sliding values (Fig. 2.21a) is:

$$y_s = D\,\sin(\frac{\pi}{2}C) \tag{2.62}$$

In order to determine an optimal guess of measured values, the initial values of the coefficients C and E from the previous expressions can be obtained using the procedure of least squares regression [27]. More often, C is given as a fixed value which is typical for the given type of curve.

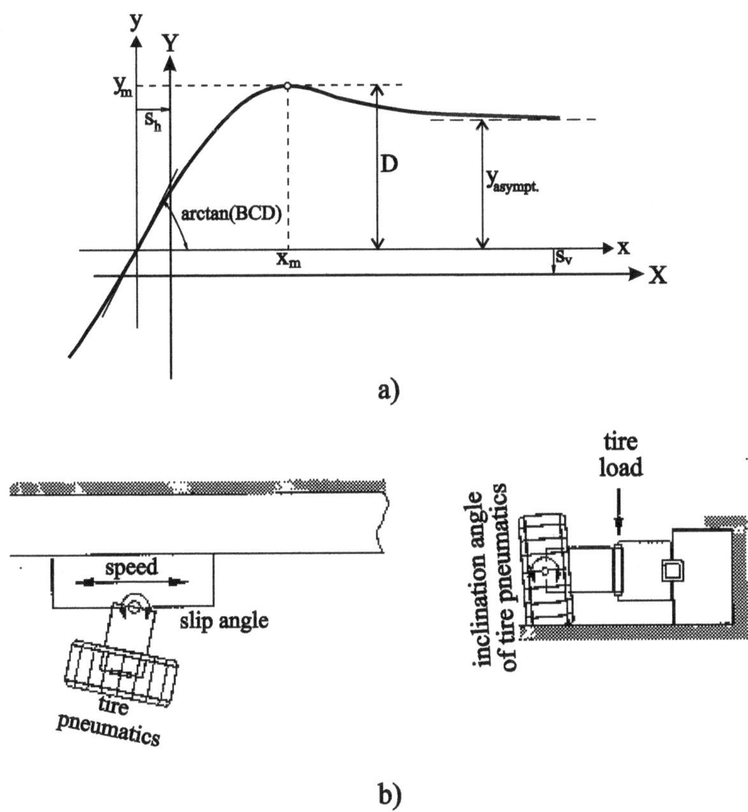

Figure 2.21: a) Typical tire characteristic describing meanings of the particular coefficients in the "magic formula", b) Slip angle and elevation angle of the tire with respect to the road surface

Additional extensions of the "magic formula" in the sense of its implementation for the cases of "combined sliding", are given in the original literature [27, 28] and [30, 31]. These references contain also instructions for the practical determination of the unknown coefficients in the model (2.58).

2.3.1.4 Open Loop Simulation of Complex Vehicle Model

In this section we shall present some selected results of numerical, open-loop simulation of a complex, spatial vehicle model whose block-diagram is shown in Fig. 2.16.

Example 2.3: The model described by the relations (2.28)-(2.54) was simulated in the open-loop by introducing control commands and external random perturbations acting on the system during its motion. The simulation results illustrate the nonlinear nature of the system model which relatively faithfully describes the real system behavior in a wide range of change of the values of state coordinates, as well as control and perturbation signals. In the considered example we neglected the effect of actuators' dynamics on total behavior of the system. The action of actuators on the behavior of the overall vehicle as a complex dynamic system will be analyzed in the subsequent section. Assuming the block-diagram of the object model shown in Fig. 2.16 some characteristic changes of the control variables δ_i and ω_i ($i = 1,\ldots,4$) were imposed as the system input. At that, the system's state coordinates as response to the given input values in a time interval were saved. Also, we stored the results of simulation of the system models describing the vehicle behavior as a reaction to the given random external disturbances of the type of road surface roughness. Variations in the road surface profile were imposed so that they act on a selected lateral pair of tires. In simulation, use was made of the vehicle model parameters given in Appendix at the end of this chapter.

In the current example, stationary vehicle motion at a constant speed of $V = 80\ [km/h]$ is given. At that, the direction of vehicle motion is imposed in two possible ways: (i) Conventional, by changing ground steering angles of the front pair of tires ($\delta_1 = \delta_2 \neq 0$) or, (ii) by changing angular velocity of front wheels. At that, the rear pair of wheels in both cases retained unchanged orientation.

In the example, numerical simulation results of which are discussed here, we consider the system response to: (1) the given trapezoidal law of change of ground steering course of motion of the front leading wheels (Fig. 2.22a); (2) given periodical sine change of the ground steering angle of the front pair of vehicle wheels (Fig. 2.22a); (3) given linear change of the front wheels ground steering angles (Fig. 2.22a), and (4) change of tire angular velocities (Fig. 2.22b). The trajectories of the vehicle body MC in particular directions in the $X - Y$ plane of road surface, as a response to the given control signals, are presented in Fig. 2.22c. In the same figure are illustrated the vehicle yaw angles as the system response to the introduced control commands shown in Fig. 2.22a and Fig. 2.22b.

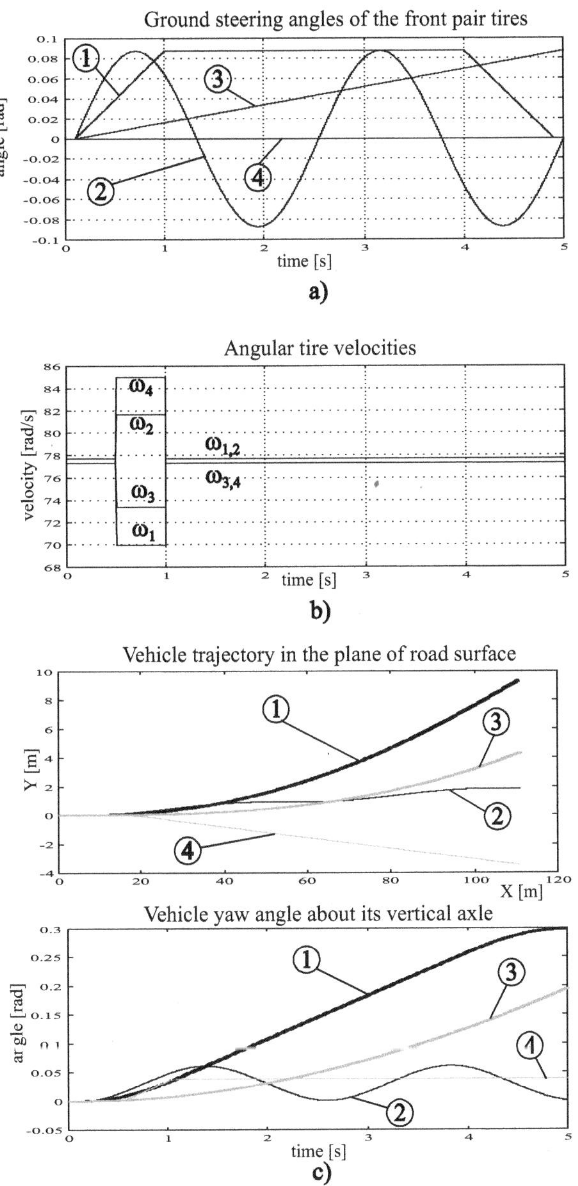

Figure 2.22: Open-loop simulation results: a) Time history of ground steering angles of front wheels, b) Time history of tire angular velocities, and c) Object responses upon the imposed perturbation

In Fig. 2.23 is illustrated the system response to the action of external disturbances of the type of a trapezoidal change of the road surface profile (road hump). In order to excite simultaneously the vehicle dynamics in the direction of rolling angle about the vehicle longitudinal symmetry axis, an asymmetric action of the road surface roughness was assigned to the lateral pair of wheels. At that, maximal value of the amplitude of road surface roughness was given $z_{r1} = z_{r3} = 9 \, [mm]$. The tire perturbation signals were delayed in time with respect to each other because of the lagging of perturbation signal of the tracking wheels with respect to the appearance of the same signal at the leading wheel. The signal time delay was given so that its value corresponded to a constant speed of vehicle motion of $V = 80 \, [km/h]$. The system's responses in the vertical Z-direction as well as in the directions of rolling ϕ and pitching θ are presented in the graphs in Fig. 2.23. By analyzing shapes of these graphs it can be found the maximal heave motion amplitude of the vehicle body MC, and it is $z \simeq 4.8$ $[mm]$. The perturbations at the lateral vehicle pair of tires cause small changes of roll angles ($\phi \simeq 0.38 \, [deg]$) and the vehicle structure pitch angle ($\theta \simeq \pm 0.08$ $[deg]$).

The results of numerical simulation presented in Figs. 2.22 and 2.23 indicate the sensitivity of the object model to the changes of particular commands as well as to the action of external disturbances that can destabilize the system in the course of its motion.

2.3.1.5 Simplified Model of Spatial Vehicle Dynamics

Simplified vehicle model can be derived from the spatial model expressed by the relations (2.28)-(2.54). By its complexity this model belongs to the group of planar vehicle models. It is presented here because it is derived from the spatial model described in detail above. The reason for describing the simplified model in this section is its suitability for the analysis of the so-called lateral dynamics of the vehicle, as well as for the synthesis of the corresponding controller of lateral dynamics [9]. Before giving details concerning the description of this vehicle model, let us adopt the following assumptions [9]:

(i) "vertical" displacement of the vehicle body MC and the structure rolling and pitching can be neglected[17],

(ii) lateral elevation of the road surface, side slip angle and deviation of the vehicle yawing angle from the corresponding nominal values are of relatively small amplitudes,

(iii) the vehicle motion speed in the longitudinal direction is constant.

Under these conditions, the above complex model (2.28)-(2.54) can be simplified and replaced with a linearized model. The number of state coordinates in the given simplified model will be four: two for motion in the lateral direction and two in the direction of vehicle yawing [9]. Referring to Fig. 2.24 [9], the derived model was defined using the following notations: y_r is the side distance

---

[17]This is in a general case an unrealistic assumption. However, it can be taken as conditionally acceptable as we assume a steady-state motion of the vehicle at a constant speed and on a plain road with no variations in its surface profile.

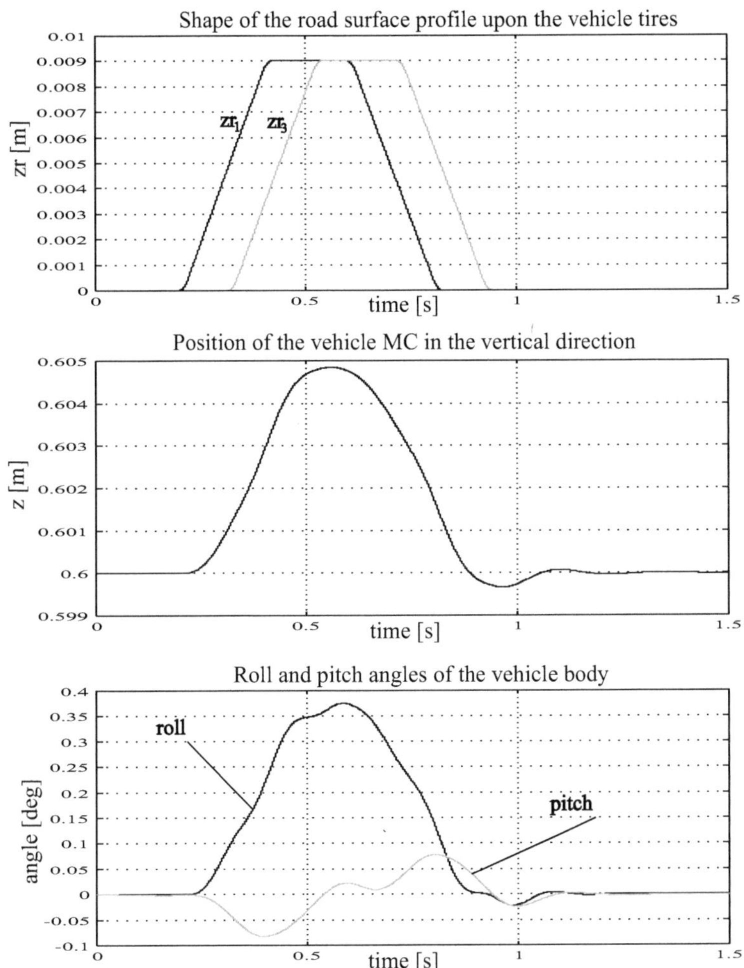

Figure 2.23: Open-loop simulation results: Response of road vehicle to variation of the road surface profile

between the vehicle MC and the road central line; ε is the vehicle body yawing angle; ε_0 is a sequence of nominal values of angles of the vehicle yawing along the given road segment; d_s is the distance of the vehicle MC from the magnetic sensor placed inside the vehicle; V is the linear vehicle speed; ρ is the radius of the road curved part; δ is the ground steering angle of the front wheels. The remaining variables used in the simplified model, whose meaning is not explicitly given here, are defined in the same way as with the complex model (2.28)-(2.54). Assuming that y_r, $\varepsilon - \varepsilon_0$ and $\dot{\varepsilon}$ are small variables, we get the relations derived from the relations (2.28)-(2.33). They represent new dynamic equations describing behavior of the simplified model [9]:

$$\ddot{y} + V\dot{\varepsilon} \approx \ddot{y}_r + \frac{V^2}{\rho}$$

$$m(\ddot{y}_r + \frac{V^2}{\rho}) = F_{wy} + \sum_{i=1}^{4} F_{B_i}$$

$$I_z \ddot{\varepsilon} = \mathcal{M}_Z \tag{2.63}$$

where F_{wy} is the lateral force of vehicle motion resistance in the y-direction and $\mathcal{M}_Z$ is the yawing moment about the vehicle vertical axis.

By its definition, the cornering stiffness C_s is:

$$C_s = \frac{\partial F_y}{\partial \alpha}\Big|_{\alpha=0} \tag{2.64}$$

It is assumed that cornering stiffness has a constant value. Also, the amplitude of lateral tire force is proportional to the value of tire side slip angle α.

$$F_y = C_s\,\alpha \tag{2.65}$$

The rate of change of the vehicle MC position with respect to the road center line is [9]:

$$\dot{y}_r = \dot{y} + V(\varepsilon - \varepsilon_0) \tag{2.66}$$

Adopting that $\dot{x} \gg s_b\dot{\varepsilon}$ we obtain from (2.65) and (2.48)

$$F_{y_i} = C_{s_i}(\delta_i - \frac{\dot{y}_r + l_1\dot{\varepsilon}}{V} + \varepsilon - \varepsilon_0), \quad i = 1, 2$$

$$F_{y_i} = C_{s_i}(\delta_i - \frac{\dot{y}_r - l_2\dot{\varepsilon}}{V} + \varepsilon - \varepsilon_0), \quad i = 3, 4 \tag{2.67}$$

Assuming that the angles in the relations (2.45) and (2.36) are small, we have:

$$\sum_{i=1}^{4} F_{B_i} = \sum_{i=1}^{4} F_{y_i} + \sum_{i=1}^{4} F_{x_i}\,\delta_i$$

$$\mathcal{M}_Z = l_1(F_{B_1} + F_{B_2}) - l_2(F_{B_3} + F_{B_4}) -$$

$$- \frac{s_b}{2}(F_{A_1} + F_{A_3}) + \frac{s_b}{2}(F_{A_2} + F_{A_4}) \tag{2.68}$$

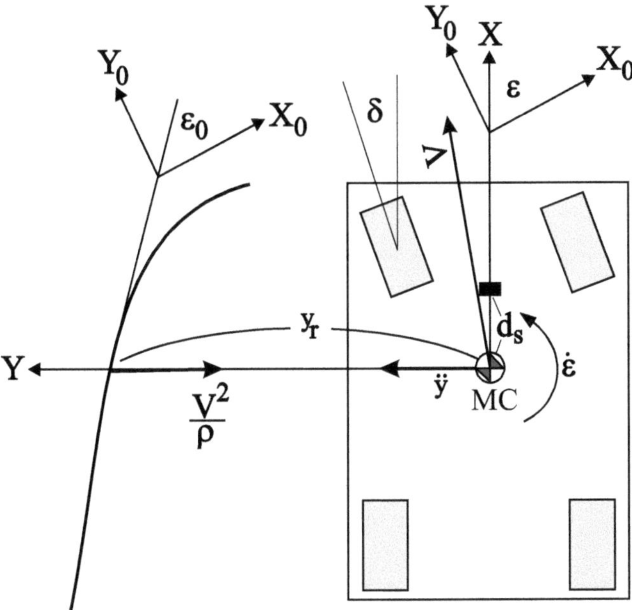

Figure 2.24: Schematic diagram of a simplified vehicle model derived from the complex tridimensional model (2.28)-(2.54)

As we consider the case of a vehicle with the possibility of changing ground steering angles on front wheels and with front-wheel drive it follows that: $\delta_1 = \delta_2 = \delta$, $\delta_3 = \delta_4 = 0$ and $F_{x3} = F_{x4} = 0$. By substituting the first of the two relations (2.68) into the middle expression of (2.63), as well as the last relation from (2.68) into the last expression of the sequence (2.63), we have:

$$\ddot{y}_r = \frac{A_1}{V}\dot{y}_r - A_1(\varepsilon - \varepsilon_0) + \frac{A_2}{V}(\dot{\varepsilon} - \dot{\varepsilon}) + B_1\delta + d_1$$

$$\varepsilon - \varepsilon_0 = \frac{A_3}{V}\dot{y}_r - A_3(\varepsilon - \varepsilon_0) + \frac{A_4}{V}(\dot{\varepsilon} - \dot{\varepsilon}) + B_2\delta + d_2 \qquad (2.69)$$

where

$$A_1 = -\frac{1}{m}[C_{s_1} + C_{s_2} + C_{s_3} + C_{s_4}]$$

$$A_2 = \frac{1}{m}[-l_1(C_{s_1} + C_{s_2}) + l_2(C_{s_3} + C_{s_4})]$$

$$A_3 = \frac{1}{I_z}[-l_1(C_{s_1} + C_{s_2}) + l_2(C_{s_3} + C_{s_4})]$$

$$A_4 \;=\; -\frac{1}{I_z}[l_1^2(C_{s_1} + C_{s_2}) + l_2^2(C_{s_3} + C_{s_4})]$$

$$B_1 \;=\; \frac{1}{m}[F_{x_1} + F_{x_2} + C_{s_1} + C_{s_2}]$$

$$B_2 \;=\; \frac{l_1}{I_z}[F_{x_1} + F_{x_2} + C_{s_1} + C_{s_2}]$$

$$d_1 \;=\; \frac{F_{wy} + F_{disturb}}{m} - \frac{V^2}{\rho} + \frac{A_2}{V}\dot{\varepsilon}_0 = \overline{d}_1 - V\dot{\varepsilon}_0 + \frac{A_2}{V}\dot{\varepsilon}_0$$

$$d_2 \;=\; \frac{T_{disturb}}{I_z} + \frac{A_4}{V}\dot{\varepsilon} - \ddot{\varepsilon} = \overline{d}_2 + \frac{A_4}{V}\dot{\varepsilon}_0 - \ddot{\varepsilon}_0 \tag{2.70}$$

The variables $F_{disturb}$ and $T_{disturb}$ have the meaning of the disturbance force and disturbance moment respectively, acting on the vehicle.

The linearized model (2.69), expressed in the state space, can finally be written in the vector form:

$$\frac{d}{dt}\begin{bmatrix} y_r \\ \dot{y}_r \\ \varepsilon - \varepsilon_0 \\ \dot{\varepsilon} - \dot{\varepsilon}_0 \end{bmatrix} = \begin{bmatrix} 0 & 1 & 0 & 0 \\ 0 & \frac{A_1}{V} & -A_1 & \frac{A_2}{V} \\ 0 & 0 & 0 & 1 \\ 0 & \frac{A_3}{V} & -A_3 & \frac{A_4}{V} \end{bmatrix} \cdot \begin{bmatrix} y_r \\ \dot{y}_r \\ \varepsilon - \varepsilon_0 \\ \dot{\varepsilon} - \dot{\varepsilon}_0 \end{bmatrix} +$$

$$+ \begin{bmatrix} 0 \\ B_1 \\ 0 \\ B_2 \end{bmatrix} \cdot \delta + \begin{bmatrix} 0 \\ d_1 \\ 0 \\ d_2 \end{bmatrix} \tag{2.71}$$

2.3.2   Typist's Chair Vehicle Model

In the discussion to follow, the conventional model presented in Fig. 2.13 is approximated by a similar model with the equivalent dynamic behavior presented in Fig. 2.25. In the equivalent model, the same vector expressions (2.28) are applied to describe the rigid body dynamics but a different approach to modeling of vehicle suspension is used. This will be explained below in this section. Instead of a 4-wheel suspension of the vehicle body we assume that the vehicle structure can relies only upon one imaginary suspension system at central position, which is in contact with the vehicle body at one system joint. In that sense, the imaginary suspension system shown in Fig. 2.25 consists of one linear and two torsional spring dampers and the corresponding viscous dampers, the values of their parameters differing from those of the conventional SS presented in Fig. 2.13. The equivalent model shown in Fig. 2.25 resembles the typist's chair, so we will call it the "typist's chair model" of road vehicle. This model is more suitable for theoretical consideration than the conventional vehicle model. Assuming the vehicle model in the "typist's chair" form, the vehicle structure can be considered as a rigid body which is supported only at one point on its "dynamic environment". In this way, all external resistances, as well as reaction forces/moments acting upon the vehicle body can be easily reduced to the MC. Such approximation is very convenient in the synthesis of dynamic control

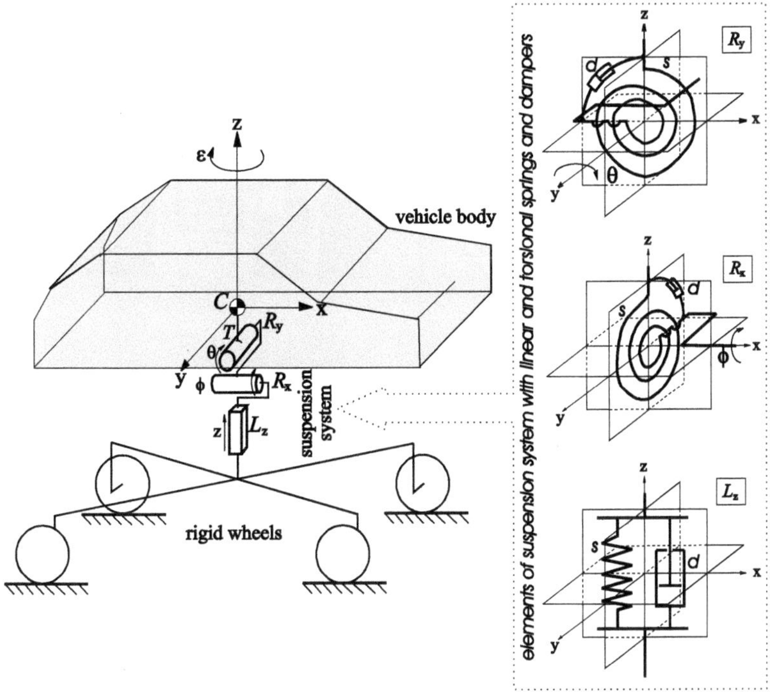

Figure 2.25: Unconventional model of the road vehicle denoted as "typist's chair" model

laws and stability analysis. Parameters of the SS presented in Fig. 2.25 can be estimated using the appropriate measurements carried out off-line. For that purpose one can use the experimental platform shown in Fig. 2.26.

In the text above we defined the notion of "dynamic environment" of the road vehicle. If we assumed that the road vehicle consists of the vehicle structure (body) and vehicle suspension with tire pneumatics, then the model of the vehicle SS in a broadest sense can approximate the vehicle environment. Thus we assume that the dynamics of tire pneumatics in "vertical" direction can be reduced to the equivalent vehicle SS dynamics. Then, it can be considered that the whole suspension's dynamic characteristics describe satisfactorily behavior of the so-called "dynamic environment" of the vehicle structure interacting with it. The "dynamic environment" model can be described in different forms. In the robotics literature, the dynamic environment is usually represented in the

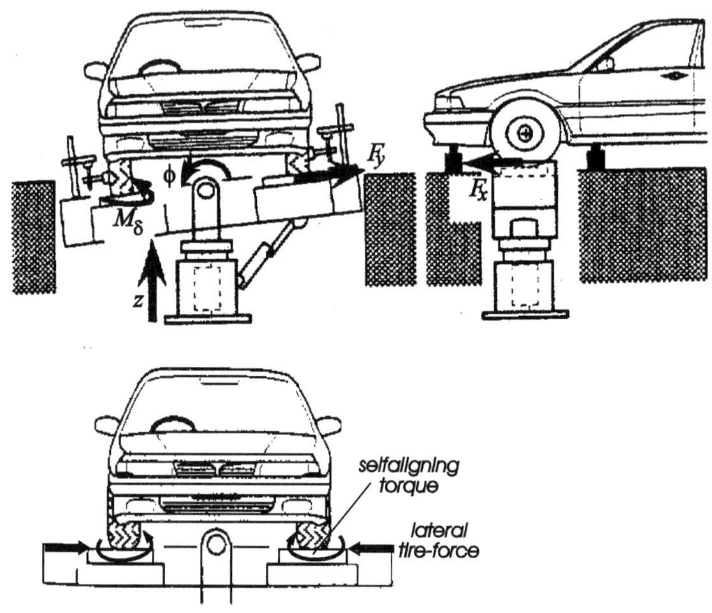

Figure 2.26: Experimental platform [1] suitable for determining parameters of the system model presented in Fig. 2.25

linear impedance form $M\ddot{q}(t) + B\dot{q}(t) + Kq(t) = -S\,F$, taking into account the inertial M, damping B, and elastic K characteristics of the environment. In some contact tasks of the robot interacting with its dynamic environment, the environment dynamics model is adopted in the nonlinear form [32]:

$$M(q,\tilde{d})\ddot{q} + L(q,\dot{q},\tilde{d}) = -S\,F \qquad (2.72)$$

where q is the (6×1) state vector, $M(q,\tilde{d})$ is a (6×6) matrix describing the equivalent environment inertia in six coordinate directions, while $L(q,\dot{q},\tilde{d})$ is the (6×1) nonlinear vector function which takes into account the equivalent elastic and damping characteristics of the environment interacting with the vehicle body. The "dynamic environment" parameters are generally variables. They are defined by the vector $\tilde{d}$ of the dimension $(\tilde{l} \times 1)$. The transformation matrix S is a (6×6) matrix which takes into account the relative orientation of the vector of external forces and moments F acting upon the environment with respect to the fixed coordinate frame attached to the ground surface.

The model of dynamic environment in the form (2.72), in accordance with the decoupling of its dynamics in particular directions, can be described by the following relation:

$$
\begin{bmatrix} M_{11} & M_{12} \\ M_{21} & M_{22} \end{bmatrix} \begin{bmatrix} \ddot{q}^{(1)} \\ \ddot{q}^{(2)} \end{bmatrix} + \begin{bmatrix} L^{(1)} \\ L^{(2)} \end{bmatrix} = -S \begin{bmatrix} F^{(1)} \\ F^{(2)} \end{bmatrix} \qquad (2.73)
$$

With the aim of simplifying our discussion we can adopt that the roll and pitch axes of the vehicle body are passing[18] through the vehicle body MC. In that case the matrix S from (2.72), i.e. (2.73), is identical to a sixth-order square unit matrix $S \equiv I$.

On the basis of experience in considering road vehicles as large-scale dynamic systems it should be pointed out that the dynamic interconnections inside the vehicle's mechanism are not of equal intensity in some particular directions. Thus, the longitudinal, lateral, and yaw motions of the vehicle body are strongly coupled. Besides, the heave motion of the vehicle body during riding is directly dependent on the corresponding displacements in the roll and pitch directions. That is the reason why the system considered (Fig. 2.25) can be dynamically decoupled in two dynamic modules: (i) the vehicle dynamics "in plane of the road surface" and, (ii) the vehicle dynamics in the "conditionally vertical plane". Dynamic coupling of the DOFs of motion inside the mentioned dynamic modules is strongly expressed, while the interactions between these modules are relatively weak. From the standpoint of previous discussion, the partitioning of the position vector q into two subvectors $q^{(1)}$ and $q^{(2)}$ is performed. The partition of the external forces vector F is carried out in the same way. As mentioned above, the vehicle body possesses $n = 6$ motion DOFs in q directions, so that the vector F of external forces and moments acting upon the vehicle structure is also of the order $m = 6$. The mentioned vectors $q^{(1)}$ and $F^{(2)}$ are of dimensions $(n_1 \times 1)$ and $(m_2 \times 1)$ respectively. The remaining two subvectors $q^{(2)}$ and $F^{(1)}$ are of dimensions $(n_2 \times 1)$ and $(m_1 \times 1)$. Having in mind all this, the vectors q and F can be defined in the following partitioned form:

$$
\begin{aligned}
q &= [q^{(1)T} \; q^{(2)T}]^T, \quad \text{where} \\
q^{(1)} &= [x \; y \; \varepsilon]^T \text{ and } q^{(2)} = [z \; \phi \; \theta \;]^T
\end{aligned} \qquad (2.74)
$$

$$
\begin{aligned}
F &= [F^{(1)T} \; F^{(2)T}]^T, \quad \text{where} \\
F^{(1)} &= [F_X \; F_Y \; M_Z]^T \text{ and } F^{(2)} = [F_Z \; M_X \; M_Y \;]^T
\end{aligned} \qquad (2.75)
$$

The elements of the described vectors belong to the set of real numbers: $q^{(1)} \in R^{n_1 \times 1}$, $q^{(2)} \in R^{n_2 \times 1}$, $F^{(1)} \in R^{m_1 \times 1}$, $F^{(2)} \in R^{m_2 \times 1}$. For the road vehicle, their dimensions are: $n = m = 6$, $n_1 = n_2 = 3$ and $m_1 = m_2 = 3$.

2.3.2.1 Parameter Estimation of Typist's Chair Vehicle Model

By analyzing the vehicle model shown in Fig. 2.25 one can percieve a specific feature of this unconventional model - the vehicle structure is supported only

---

[18]In a general case, these axes do not pass exactly through the vehicle body MC, but they are somewhat shifted.

at one point (denoted by T), making a unique SS. In this case the SS system consists of three springs and three viscous dampers. The symbols L_z, R_x and R_y denote the possible DOFs of the vehicle body: one translational and two rotational directions. In these three DOF directions are placed one linear and two torsional springs, as well as the corresponding viscous shock absorbers (see Fig. 2.25). Through these elements, the disturbances caused by the unevenness of the road surface are transmitted to the point T, which represents the suspension joint center (Fig. 2.25). In accordance with the previously given definition of the notion of environmental dynamics, the problem of estimating parameters of the model of this environment is reduced to the determination of the corresponding stiffness coefficients of the springs and damping coefficients of the viscous dampers presented in Fig. 2.25. The stiffness coefficients of the linear and torsional springs in the model shown in Fig. 2.25, as well as the corresponding damping coefficients of viscous dampers are determined by one of the known methods. In the simulation example to be presented below we will give the way of practical estimation of the unknown parameters of the environment model (2.73) in the frame of the unconventional typist's chair vehicle model.

Example 2.4:   Estimation of the vehicle environment model can be carried out in three steps:

(i) In the first step we determine static characteristics of the equivalent SS whose model is presented in Fig. 2.25. They are determined on the basis of the results obtained by measurements carried out on a real system. For this purpose there are nowadays different types of the so-called testing stations [1, 33]. One type of experimental testing station is constructed in the form of a hydraulic platform [1]. It enables the determination of both static and dynamic characteristics of the vehicle motion in particular coordinate directions under laboratory conditions. Such test platform is presented in Fig. 2.26 [1].

In the current example, instead of the measurement results obtained on a real system use was made of the results obtained by simulation of the nonlinear spatial model described by the relations (2.28)-(2.54). The system presented in Fig. 2.13 is subjected to slow static loading by giving forces and moments at the vehicle body MC. First, we apply the force F_Z acting at the vehicle body MC in the vertical direction perpendicular to the ground plane. This force causes MC displacements in the direction of the Z-axis. When the action of the force F_Z comes to an end, the moment M_X acting at the vehicle body MC about its longitudinal X-axis is given. This moment produces the corresponding angular displacements of the vehicle body for the rolling angle ϕ. Then, the system is perturbed with the pitch moment M_Y at MC, about the transversal rotation axis, causing the structure rotation about the Y-axis by the angle θ. In the simulation example, amplitudes of the mentioned forces and moments were gradually changed in the range of their real values. At the same time, values of the corresponding changes in position and orientation of the vehicle body observed in the particular coordinate directions are saved. In this way, the following pairs of functional relationships were obtained: $F_Z - z$, $F_Z - \phi$,

$F_Z - \theta$, $M_X - z$, $M_X - \phi$, $M_X - \theta$, $M_Y - z$, $M_Y - \phi$ and $M_Y - \theta$. They represent static characteristics of the model presented in Fig. 2.13. Since the system model shown in Fig. 2.25, on the basis of its characteristics, should exhibit the dynamic behavior equivalent to that of the model illustrated in Fig. 2.13, this implies that static characteristics of the system obtained by simulation of the system from Fig. 2.13 should be equal to the characteristics of the model shown in Fig. 2.25, whose parameters we want to estimate.

In Fig. 2.27 are presented the functional curves obtained by numerical simulation of the model (2.28)-(2.54) from Fig. 2.13. They were used to estimate the environment model parameters in the expression (2.73). The values of geometrical and dynamic parameters of the vehicle model shown in Fig. 2.13 and the model used in this simulation experiment are given in Appendix at the end of this chapter.

(ii) In the second step of the procedure for estimating the environment model parameters appropriate mathematical approximation was carried out. The obtained functional dependence curves (Fig. 2.27) are fitted by higher-order polynomials. In the current example use was made of the fifth-degree polynomials. The polynomial coefficients describe the nonlinear nature of behavior of the springs of the SS presented in Fig. 2.25. The chosen approximation polynomials are of the following form:

$$F_Z = \sum k_{ij}^{(n)} z^n, \quad i = 1, \ j = 1, \ldots, 3, \ n = 1, \ldots, 5$$

$$M_X = \sum k_{ij}^{(n)} \phi^n, \quad i = 2, \ j = 1, \ldots, 3, \ n = 1, \ldots, 5$$

$$M_Y = \sum k_{ij}^{(n)} \theta^n, \quad i = 3, \ j = 1, \ldots, 3, \ n = 1, \ldots, 5 \tag{2.76}$$

Values of the parameters $k_{ij}^{(n)}$, obtained by fitting of the curves in Fig. 2.27, representing the SS static characteristics, are given in Table 2.2.

(iii) In this step we determined the inertia parameters of the environment model in the expression (2.73) as well as the corresponding damping coefficients of viscous dampers in the term $L^{(2)}$ of the model (2.73). The system being in the state of relative rest was perturbed by a random time-dependent signal of different amplitude and frequency acting on each particular tire. The given signals were pseudo-random functions of amplitude in the range of $0 - 15$ $[mm]$ and frequency in the range of $5 - 25$ $[Hz]$. The signal frequency of 25 $[Hz]$ corresponds to encountering a hump or a dip on the road on each ≈ 1 $[m]$ at a constant riding speed of 80 $[km/h]$. Samples of simulation results were recorded each 2 $[ms]$. The state variables involved were: vehicle body positions in the respective directions z, ϕ, θ, the corresponding changes of velocities ($\dot{z}$, $\dot{\phi}$, $\dot{\theta}$) and accelerations ($\ddot{z}$, $\ddot{\phi}$, $\ddot{\theta}$). At the same time, the corresponding forces and moments (F_Z, M_X, M_Y) caused by the changes of the mentioned state variables, are saved. The results of simulation thus obtained were grouped into the corresponding vectors which are used for an off-line determination of the environment parameters as was described in [34]. For this purpose we assumed first the mathematical form of the model structure of the "environment" whose

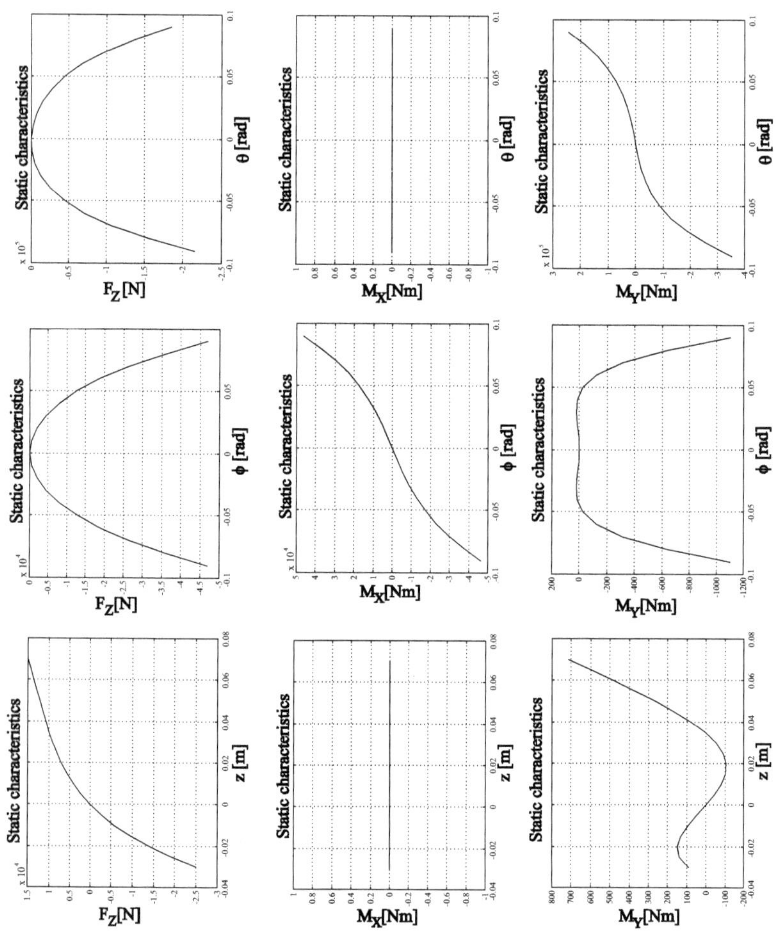

Figure 2.27: Curves of functional dependence of the system displacements on the corresponding loads at the vehicle body MC, obtained by the simulation experiment

$\overline{m}_{ij}$	j=1	j=2	j=3
i=1	-894.19	-19.08	6.57
i=2	0.00	-237.62	0.00
i=3	38.37	43.94	-1260.60

$\underline{m}_{ij}$	j=1	j=2	j=3
i=1	0.0005	-0.1344	0.1805
i=2	-0.0166	0.1161	0.8648
i=3	-0.0225	-0.7936	1.6450

$\overline{b}_{ij}$	j=1	j=2	j=3
i=1	11953.00	-186.20	-6288.70
i=2	0.00	7015.30	0.00
i=3	25108.00	-1267.20	12956.00

$\underline{b}_{ij}$	j=1	j=2	j=3
i=1	0.0001	-0.0948	1.7425
i=2	-0.0074	1.4053	-5.5330
i=3	-0.0064	0.3828	7.3047

$k_{ij}^{(n)}$		n=1	n=2	n=3	n=4	n=5
i=1	j=1	$4.8374*10^{5}$	$-8.5117*10^{6}$	$9.4691*10^{7}$	$-3.9990*10^{8}$	$-4.8784*10^{8}$
	j=2	$2.5211*10^{-11}$	$-4.7816*10^{6}$	$-2.5604*10^{-8}$	$-1.2777*10^{8}$	$2.8673*10^{-6}$
	j=3	$-5.4749*10^{4}$	$-1.5255*10^{7}$	$2.0139*10^{7}$	$-1.1872*10^{9}$	$8.6788*10^{8}$
i=2	j=1	0	0	0	0	0
	j=2	$2.7215*10^{5}$	$7.7306*10^{-11}$	$2.9899*10^{7}$	$7.0385*10^{-8}$	$2.3835*10^{6}$
	j=3	0	0	0	0	0
i=3	j=1	$-9.6712*10^{3}$	$1.0199*10^{5}$	$7.9971*10^{6}$	$-9.0050*10^{7}$	$1.8402*10^{8}$
	j=2	$1.0393*10^{-12}$	$4.6702*10^{4}$	$-5.2133*10^{-10}$	$-2.2698*10^{7}$	$4.1938*10^{-8}$
	j=3	$8.4722*10^{5}$	$-2.6263*10^{6}$	$2.8922*10^{8}$	$-5.2686*10^{8}$	$2.2223*10^{9}$

Table 2.2: Values of the parameters of the environment dynamics model (2.73) obtained in the simulation example

parameters are sought. This can be expressed by the following relations:

$$F_Z = \overline{m}_{11}\ddot{z} + \overline{m}_{12}\ddot{\phi} + \overline{m}_{13}\ddot{\theta} + \overline{b}_{11}\dot{z} + \overline{b}_{12}\dot{\phi} + \overline{b}_{13}\dot{\theta} + \tag{2.77}$$
$$+ \; k_{11}^{(1)}z + \ldots + k_{11}^{(5)}z^5 + k_{12}^{(1)}\phi + \ldots + k_{12}^{(5)}\phi^5 + k_{13}^{(1)}\theta + \ldots + k_{13}^{(5)}\theta^5$$

$$M_X = \overline{m}_{21}\ddot{z} + \overline{m}_{22}\ddot{\phi} + \overline{m}_{23}\ddot{\theta} + \overline{b}_{21}\dot{z} + \overline{b}_{22}\dot{\phi} + \overline{b}_{23}\dot{\theta} + \tag{2.78}$$
$$+ \; k_{21}^{(1)}z + \ldots + k_{21}^{(5)}z^5 + k_{22}^{(1)}\phi + \ldots + k_{22}^{(5)}\phi^5 + k_{23}^{(1)}\theta + \ldots + k_{23}^{(5)}\theta^5$$

$$M_Y = \overline{m}_{31}\ddot{z} + \overline{m}_{32}\ddot{\phi} + \overline{m}_{33}\ddot{\theta} + \overline{b}_{31}\dot{z} + \overline{b}_{32}\dot{\phi} + \overline{b}_{33}\dot{\theta} + \tag{2.79}$$
$$+ \; k_{31}^{(1)}z + \ldots + k_{31}^{(5)}z^5 + k_{32}^{(1)}\phi + \ldots + k_{32}^{(5)}\phi^5 + k_{33}^{(1)}\theta + \ldots + k_{33}^{(5)}\theta^5$$

where

$$M_{22} = \begin{bmatrix} \overline{m}_{11} & \overline{m}_{12} & \overline{m}_{13} \\ \overline{m}_{21} & \overline{m}_{22} & \overline{m}_{23} \\ \overline{m}_{31} & \overline{m}_{32} & \overline{m}_{33} \end{bmatrix}; \quad B_2 = \begin{bmatrix} \overline{b}_{11} & \overline{b}_{12} & \overline{b}_{13} \\ \overline{b}_{21} & \overline{b}_{22} & \overline{b}_{23} \\ \overline{b}_{31} & \overline{b}_{32} & \overline{b}_{33} \end{bmatrix}; \tag{2.80}$$

The model represented by the equations (2.77)-(2.79) can be written in a linear vector from [34]:

$$\mathcal{Y} = \mathcal{U}\Theta + e \tag{2.81}$$

where, $\mathcal{Y}$ is the $(n_e \times 1)$ output vector; $\mathcal{U}$ is the $(n_e \times m_e)$ input vector; e is the $(n_e \times 1)$ vector of model error; Θ is the vector of model parameters to be determined. The symbol n_e denotes the number of equidistant time samples in the estimated vector, whereas m_e represents the number of model parameters in the vector Θ. The equation (2.81) should be solved for Θ applying pseudo-inversion in the way described by the expression:

$$\Theta = [\mathcal{U}^T\mathcal{U}]^{-1}\mathcal{U}^T\mathcal{Y} \tag{2.82}$$

The described procedure is known as the "ordinary least squares method" [34]. By the model (2.77)-(2.79), the vectors $\mathcal{Y}, \mathcal{U}$ and Θ have the following elements:

$$\mathcal{Y} = F_Z - (k_{11}^{(1)}z + \ldots + k_{11}^{(5)}z^5 + k_{12}^{(1)}\phi + \ldots +$$
$$+ \; k_{12}^{(5)}\phi^5 + k_{13}^{(1)}\theta + \ldots + k_{13}^{(5)}\theta^5)$$
$$\mathcal{U} = [\ddot{z} \;\; \ddot{\phi} \;\; \ddot{\theta} \;\; \dot{z} \;\; \dot{\phi} \;\; \dot{\theta}]$$
$$\Theta = [\overline{m}_{11} \;\; \overline{m}_{12} \;\; \overline{m}_{13} \;\; \overline{b}_{11} \;\; \overline{b}_{12} \;\; \overline{b}_{13}]^T \tag{2.83}$$

$$\mathcal{Y} = M_X - (k_{21}^{(1)}z + \ldots + k_{21}^{(5)}z^5 + k_{22}^{(1)}\phi + \ldots +$$
$$+ \; k_{22}^{(5)}\phi^5 + k_{23}^{(1)}\theta + \ldots + k_{23}^{(5)}\theta^5)$$
$$\mathcal{U} = [\ddot{z} \;\; \ddot{\phi} \;\; \ddot{\theta} \;\; \dot{z} \;\; \dot{\phi} \;\; \dot{\theta}]$$
$$\Theta = [\overline{m}_{21} \;\; \overline{m}_{22} \;\; \overline{m}_{23} \;\; \overline{b}_{21} \;\; \overline{b}_{22} \;\; \overline{b}_{23}]^T \tag{2.84}$$

$$\mathcal{Y} = M_Y - (k_{31}^{(1)}z + \ldots + k_{31}^{(5)}z^5 + k_{32}^{(1)}\phi + \ldots +$$
$$+ \; k_{32}^{(5)}\phi^5 + k_{33}^{(1)}\theta + \ldots + k_{33}^{(5)}\theta^5)$$
$$\mathcal{U} = [\ddot{z} \;\; \ddot{\phi} \;\; \ddot{\theta} \;\; \dot{z} \;\; \dot{\phi} \;\; \dot{\theta}]$$
$$\Theta = [\overline{m}_{31} \;\; \overline{m}_{32} \;\; \overline{m}_{33} \;\; \overline{b}_{31} \;\; \overline{b}_{32} \;\; \overline{b}_{33}]^T \tag{2.85}$$

Values of the parameters $\overline{m}_{ij}$ forming the inertia matrix M_{22} in (2.73) and the values of the parameters $\overline{b}_{ij}$ forming the matrix of damping B_2 are given in Table 2.2, too.

In the course of riding at higher speeds and variable accelerations, abrupt change of the motion direction, sudden braking, cornering, etc., inertia effects in the system come into effect. In addition to the characteristic effects observed in the longitudinal and lateral directions with respect to the longitudinal axis of the vehicle, the mentioned causes produce relative changes in the position of vehicle structure with respect to road surface plane observed as heave, pitch, and roll motion. As a result, certain forces and moments (F_Z, M_X, M_Y) arise at the vehicle body MC. They indirectly influence the system's dynamic behavior not only in the plane perpendicular to the road surface but also in the plane of vehicle motion (longitudinal and lateral directions and direction of yawing about the vertical axis of the vehicle). In order to determine the corresponding parameters in the environment model (2.73), related to the vehicle dynamics in the road surface plane, a procedure similar to that described above to deal with the effects of the mentioned forces and moments on motion in the z, ϕ i θ directions of the vertical plane, is applied. Starting again from the spatial complex model shown in Fig. 2.13, the corresponding amplitudes of changes of the motion velocity and direction are given to perturb the planar dynamics of the vehicle. At that, the values of velocities $\dot{x}$, $\dot{y}$, $\dot{\varepsilon}$, accelerations $\ddot{x}$, $\ddot{y}$, $\ddot{\varepsilon}$, and the corresponding forces and moments F_z, M_x, M_y appearing at the vehicle body MC as a consequence of the input changes, are saved. The stored values of these variables, obtained by numeric simulation of the model (2.28), are grouped again into the corresponding vectors in a similar way as described above. In this case, the vectors $\mathcal{Y}$, $\mathcal{U}$ and Θ have the following elements:

$$
\begin{aligned}
\mathcal{Y} &= F_Z - F_Z^r \\
\mathcal{U} &= [\ddot{x}\ \ddot{y}\ \ddot{\varepsilon}\ \dot{x}\ \dot{y}\ \dot{\varepsilon}] \\
\Theta &= [\underline{m}_{11}\ \ \underline{m}_{12}\ \ \underline{m}_{13}\ \ \underline{b}_{11}\ \ \underline{b}_{12}\ \ \underline{b}_{13}]^T
\end{aligned}
\tag{2.86}
$$

$$
\begin{aligned}
\mathcal{Y} &= M_X \\
\mathcal{U} &= [\ddot{x}\ \ddot{y}\ \ddot{\varepsilon}\ \dot{x}\ \dot{y}\ \dot{\varepsilon}] \\
\Theta &= [\underline{m}_{21}\ \ \underline{m}_{22}\ \ \underline{m}_{23}\ \ \underline{b}_{21}\ \ \underline{b}_{22}\ \ \underline{b}_{23}]^T
\end{aligned}
\tag{2.87}
$$

$$
\begin{aligned}
\mathcal{Y} &= M_Y \\
\mathcal{U} &= [\ddot{x}\ \ddot{y}\ \ddot{\varepsilon}\ \dot{x}\ \dot{y}\ \dot{\varepsilon}] \\
\Theta &= [\underline{m}_{31}\ \ \underline{m}_{32}\ \ \underline{m}_{33}\ \ \underline{b}_{31}\ \ \underline{b}_{32}\ \ \underline{b}_{33}]^T
\end{aligned}
\tag{2.88}
$$

where F_Z^r is the equilibrium value of the vehicle payload F_Z, which is equal to the vehicle gravity force, $F_Z^r = m \cdot g$.

The elements $\underline{m}_{ij}$ and $\underline{b}_{ij}$ form the matrix of inertia M_{21} and the matrix of damping B_1 make the matrices:

$$M_{21} = \begin{bmatrix} \underline{m}_{11} & \underline{m}_{12} & \underline{m}_{13} \\ \underline{m}_{21} & \underline{m}_{22} & \underline{m}_{23} \\ \underline{m}_{31} & \underline{m}_{32} & \underline{m}_{33} \end{bmatrix} ; \quad B_1 = \begin{bmatrix} \underline{b}_{11} & \underline{b}_{12} & \underline{b}_{13} \\ \underline{b}_{21} & \underline{b}_{22} & \underline{b}_{23} \\ \underline{b}_{31} & \underline{b}_{32} & \underline{b}_{33} \end{bmatrix} ; \quad (2.89)$$

The values of the parameters $\underline{m}_{ij}$ and $\underline{b}_{ij}$ obtained in the current simulation example are also listed in Table 2.2.

After estimating all the relevant parameters of the environment model from the relation (2.73) it is possible to write down a vector relation making the basis for determining values of the unknown forces and moments F_Z, M_X and M_Y acting at the vehicle body MC. Also, the condition that the corresponding system motion variables (positions, velocities, and accelerations) are known in particular directions, must be satisfied. Such a relation can be now written in the form:

$$F^{(2)} = M_{21}\ddot{q}^{(1)} + M_{22}\ddot{q}^{(2)} + L^{(2)} = [F_Z \ M_X \ M_Y]^T \quad (2.90)$$

$$L^{(2)}(q,\dot{q}) = B_1\dot{q}^{(1)} + B_2\dot{q}^{(2)} + \begin{bmatrix} s_{11} + s_{12} + s_{13} \\ s_{21} + s_{22} + s_{23} \\ s_{31} + s_{32} + s_{33} \end{bmatrix} \quad (2.91)$$

where s_{ij} are the forces and moments due to the springs of the SS shown in Fig. 2.25, and the terms $B_1\dot{q}^{(1)}$ and $B_2\dot{q}^{(2)}$ represent the corresponding damping forces and moments in the viscous dampers. Amplitudes of the forces/moments s_{ij} in the preceding expression are calculated from the following scalar relations:

$$\begin{aligned}
s_{i1} &= k_{i1}^{(1)} z + \ldots + k_{i1}^{(5)} z^5, & i = 1,\ldots,3 \\
s_{i2} &= k_{i2}^{(1)} \phi + \ldots + k_{i2}^{(5)} \phi^5, & i = 1,\ldots,3 \\
s_{i3} &= k_{i3}^{(1)} \theta + \ldots + k_{i3}^{(5)} \theta^5, & i = 1,\ldots,3
\end{aligned} \quad (2.92)$$

Example 2.5:   In this example we present the comparative results of numerical simulation involving two cases: (i) the case of implementation of complex spatial vehicle model shown in Fig. 2.13, and (ii) the case of implementation of the equivalent typist's chair vehicle model presented in Fig. 2.25. Simulation experiments were carried out for the same vehicle trajectory and motion conditions. At that, use was made of the estimated values of parameters of the environment model given in Table 2.2, which were determined by the procedure described in Example 2.4. Arbitrary curvilinear motion of the vehicle was given, as well as stochastic disturbances of the type of change in road surface profile. These disturbances of variable amplitudes act independently on each vehicle tire. At that, maximum amplitude of surface road profile variation was 8 [mm]. Comparative results were shown in Fig. 2.28. By comparing results of the realized amplitudes of forces and moments $F_Z(t)$, $M_X(t)$ and $M_Y(t)$ acting at the vehicle body MC for the same given trajectory and motion conditions, the following conclusion can be drawn. The environment model parameters taken from Table 2.2, determined in the way described in simulation Example 2.4,

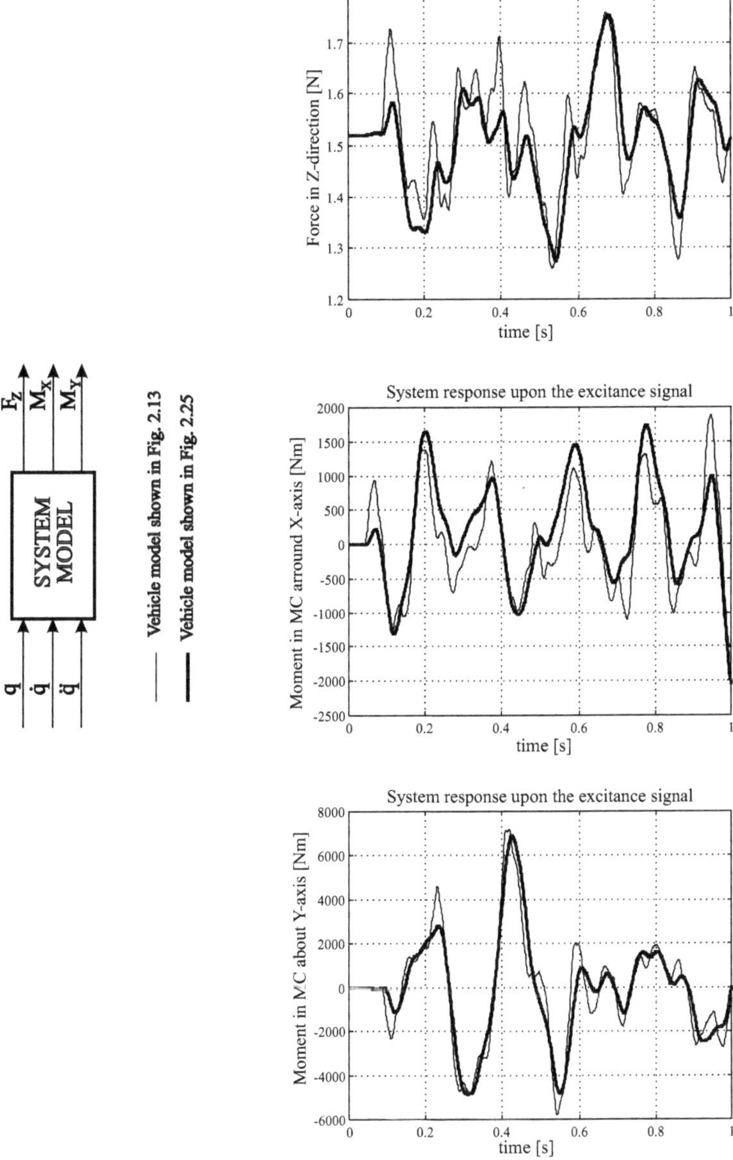

Figure 2.28: Comparative simulation results of the convential vehicle model (Fig. 2.13) and the typist's chair model (Fig. 2.25) with the equivalent dynamic behavior

ensure a relatively good similarity of dynamic behavior of the vehicle models given in Fig. 2.13 and Fig. 2.25. It is thus shown that the conventional spatial model given by the relations (2.28)-(2.54) may be practically replaced by the corresponding model of the equivalent behavior which, as will be demonstrated later, offers certain advantages when the synthesis of dynamic control laws is concerned. Of course, the degree of accuracy of estimation of the environment model parameters (2.73) is of great importance in the synthesis of the vehicle dynamics control. This fact has to be born in mind when we want to design a robust vehicle control with satisfactory accuracy.

2.3.3 Integrated Model of Vehicle Dynamics

The overall behavior of a vehicle is influenced not only by the dynamic and geometrical parameters of its structure, characteristics of SS and tire pneumatics, but also by performances of the actuators of its driving-steering subsystem. Contemporary road vehicles possess different types of actuators (driving units) operating within the so-called active systems such as active suspension system, active steering system, active braking/traction system, etc. Nowadays, the actuators that are most widely used with road vehicles can be classified according to their nature as mechanical, electrical, hydraulic or combined (electromechanical, electrohydraulic, etc.).

Under the "integrated vehicle model" we assume the model which by its mathematical expressions unites the descriptions of dynamic behavior of the vehicle structure together with its suspension system and tire pneumatics as well as with the dynamics of the executive organs - actuators. The integrated model describes the overall dynamics of a road vehicle together with its driving-steering subsystems. This model is derived by functionally connecting the relations of rigid body dynamics of the vehicle structure, model of the elasto-dynamics of SS and tire pneumatics, and the relations describing the dynamics of the corresponding actuators. In this section we will present an example of deriving the integrated model of an automated road vehicle with the actuators selected on the basis of the existing designs employed in the industrial practice [35, 36], [37, 38], [39].

2.3.3.1 Actuators and Active Control Systems with Road Vehicles

In Fig. 2.29 is presented an electrohydraulic actuator which is used as an active damping cylinder within the vehicle ASS. The active hydro-cylinder generates an active damping force on its piston, by which it influences the ASV performance in the vertical direction [35, 36].

In Fig. 2.30 is shown the functional scheme of an active four-wheel steering (A4WS) system. An automated road vehicle should have the possibility of changing independently the ground steering angle on all four wheels (independent active wheel steering). By applying I4WS the maneuvering capabilities of the vehicle are improved as well as the controllability of its lateral dynamics. The presented functional scheme is based on the design solutions given in [37, 38].

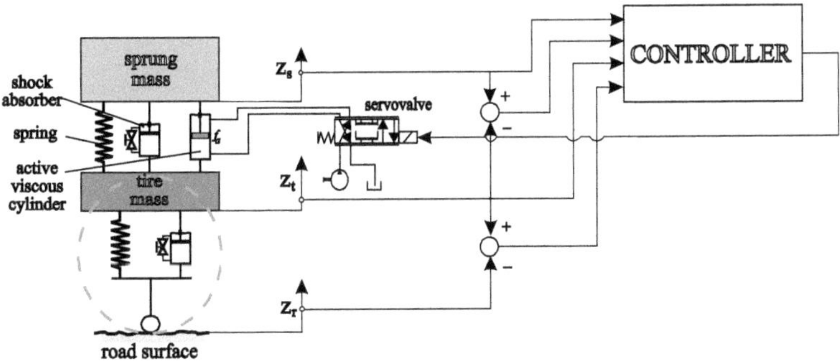

Figure 2.29: Functional-symbolic scheme of an active suspension system with active damping cylinder, servovalve and local controller

The hydraulic cylinder force F_{δ_i} (Fig. 2.30), with the aid of the leverage, produces wheel displacement about its vertical rotation axis. At that, each wheel can realize independently the given orientation with respect to the direction of the vehicle longitudinal symmetry axis. The hydro-cylinder is microprocessor-controlled, with the aid of an electromechanical servovalve as positional device.

In Fig. 2.31a is presented one of unconventional vehicles with hybrid drivetrain. The presented driving concept is known as "hybrid power drivetrain" [39]. Such power-driving system appeared to be a good alternative to conventional mode which is based upon the use of internal combustion engine and the corresponding hydromechanical transmission of power to the vehicle wheels. In contrast to conventional cars, the driving power of the internal combustion engine of the "hybrid vehicle" is used to power an electric current generator (Fig. 2.31b). The obtained voltage is microprocessor-modulated and fed to an electric motor, which directly produces rotational motion of the vehicle wheels. In Fig. 2.31a are shown four electric motors, one on each vehicle wheel. The driving electric motors are controlled by microprocessors in a way which depends on the concrete type of actuator (AC or DC motor). In Fig. 2.31b [39] is schematically presented the way of controlling the angular velocity ω_i of rotation of the i-th vehicle wheel. The advantage of this hybrid drive train is that it enables direct control of the rotation speed of each wheel. In that way a better controllability of the system is attained. Control of angular velocity of each wheel can actively change the tire forces and moments, which yields enhanced vehicle stability in both longitudinal and lateral directions. Synchronized control of the 4-tire angular velocities allows a direct control of the moment of the vehicle body yawing about its vertical axis which is not possible with conventional vehicles. Combined action of the vehicle active traction system (hybrid drive train) and active

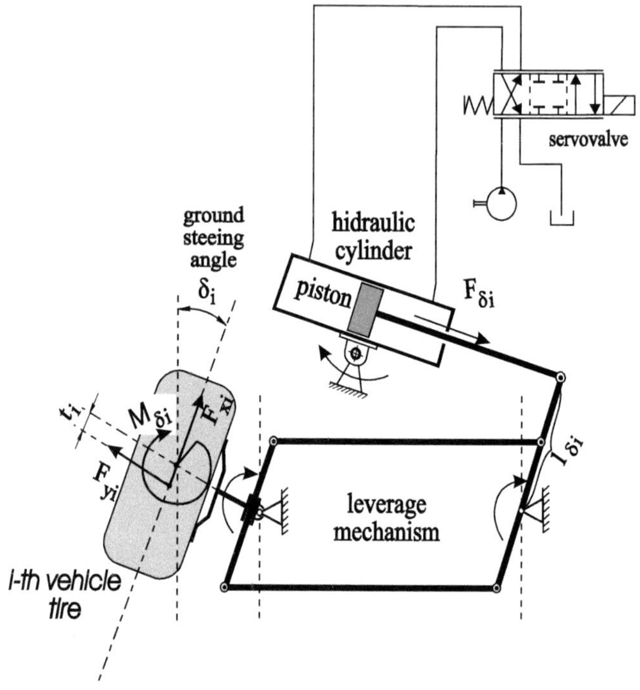

Figure 2.30: Functional-symbolic scheme of active wheel steering system involving hydraulic cylinder and servovalve

system of steering-braking yield an essential improvement of the vehicle maneuverability and controllability in both stationary regime and transition regimes of motion.

Generally speaking, an automated road vehicle represents an active dynamic system whose particular DOFs of motion are powered with actuators. The vehicle actuators actively change the vehicle dynamic behavior. In this way they enable its adaptation to the changeable conditions of the environment. Cocerning that, the outer disturbances acting most often on the system and tending to destabilize it are: the effect of road surface unevenness, effect of sudden side wind gust, effect of variation of the coefficient of tire-road friction, etc. On the other hand, internal disturbances acting on the system are usually a consequence of stochastic changes in the system itself. As example we can mention the pressure drop in the tire pneumatics caused by their damage, change in the mass and moment of inertia caused by displacement of the vehicle load or,

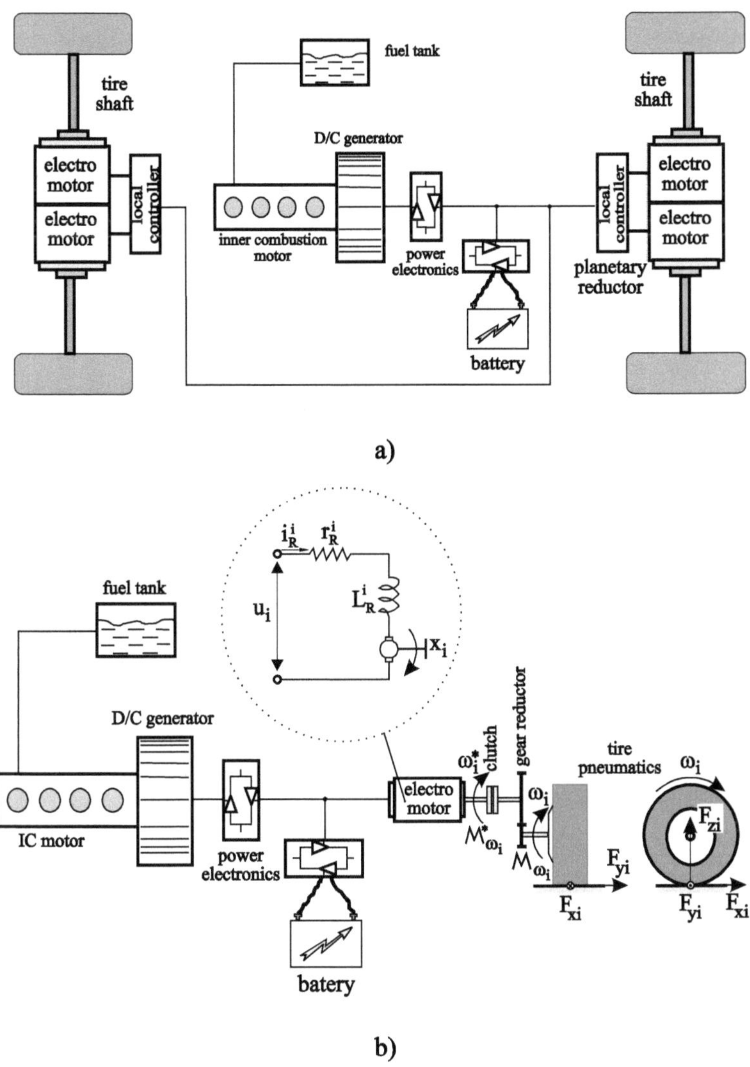

Figure 2.31: a) Schematic of a hybrid power drivetrain, b) Functional-symbolic scheme of active four-wheel system of driving/braking using an electric motor

its falling out, etc. Because of the great importance of actuators in control of the vehicle dynamic performance it is useful to introduce into the vehicle model the actuators' dynamics, to allow their simulation and, thus the integrated system analysis.

2.3.3.2 General Model of Vehicle Actuator

General mathematical model of an actuator irrespective of its type (mechanical, hydraulic, electrical) will be derived in the text to follow. The general actuator model is usually nonlinear and is expressed in state space. As a rule, its order is higher than 2. The equations of state and actuator output equation in its vector form are given by the following relations [42, 43]:

$$
\begin{aligned}
\dot{x}_i &= g_i(x_i, p_i) + b_i(p_i)N(u_i) + f_i(p_i)F_{o_i} \\
y_i &= G_i x_i
\end{aligned}
\tag{2.93}
$$

where x_i is the $(n_a \times 1)$ system state vector; p_i is the vector of the i-th actuator parameters; u_i is the scalar value of the input (control) signal of the i-th actuator; F_{o_i} is the scalar value of the force/load of the actuator output element[19]; y_i is the system's output variable; G_i is the system ourput matrix. The values of the control signals u_i are usually of restricted amplitude and can be described by the nonlinearity of saturation type. By their nature, the control variables depend on the chosen type of actuators and they are most often electrical signals (voltage or current). The presented general model of actuator (2.93) is needed to derive the integrated vehicle model.

2.3.3.3 Relations Describing Integrated Vehicle Model

Let us assume that the full control of an automated road vehicle can be realized with the aid of 12 selected actuators. On each wheel are installed three actuators: (i) active damping cylinder for the active suspension, (ii) hydraulic cylinder for active wheel steering, and (iii) electric motor for regulating the wheel angular velocity. In our consideration we assume that each actuator can function independently of each other. This is practically a most complex case of application. In practice, there are solutions in which particular actuators have a dual function, say for example an actuator can control the orientation of one pair (the front or rear) of tires [37].

To form an integrated vehicle model it is necessary to establish mathematical relations between the vectors of generalized forces and moments τ acting at the vehicle body MC and the corresponding forces and moments $f_{a_i}, F_{\delta i}, \mathcal{M}_{\omega_i}$ at the actuator output elements (see Figs. 2.29, 2.30 and 2.31). The vector of generalized forces and moments τ in the model (2.28) is determined on the basis of the knowledge of forces at the SS joints. The longitudinal F_{A_i}, lateral F_{B_i}, and vertical P_i forces at the joints (see Fig. 2.13) produce the moments

---

[19]With a hydraulic cylinder and electrical motor the output elements are the piston and shaft.

of rolling $\mathcal{M}_X$, pitching $\mathcal{M}_Y$ and yawing $\mathcal{M}_Z$ of the vehicle body about its main inertia axes. The vector of generalized forces τ is defined by the relation (2.33). The moments $\mathcal{M}_X$, $\mathcal{M}_Y$ and $\mathcal{M}_Z$ are consequences of the existence of the mentioned forces. These moments are calculated from the relations (2.34)-(2.36). The vertical force P_i, longitudinal force F_{A_i} and lateral force F_{B_i} at the i-th vehicle SS joint are determined from the relations (2.42) and (2.45). Functional relation between the longitudinal interaction force F_{x_i} of the i-th tire and the load moment at the output shaft of the electrical motor (Fig. 2.31b) is given as:

$$F_{x_i} = \frac{N_m^i \mathcal{M}_{\omega_i}^*}{r_i} = \frac{\mathcal{M}_{\omega_i}}{r_i} \tag{2.94}$$

where N_m^i is the reduction gear ratio of the moment from the output to input axle of the i-th actuator reductor; $\mathcal{M}_{\omega_i}^*$ is the load moment at the electric motor output; $\mathcal{M}_{\omega_i}$ is the load moment at the reductor input axle; r_i is the effective radius of the pneumatic which depends on the vertical tire load. The relationship providing the kinematic connection between the angular velocity of the i-th wheel and angular velocity of rotation of the corresponding motor shaft is:

$$\omega_i = N_v^i \omega_i^* \tag{2.95}$$

where ω_i and ω_i^* are respective angular velocities of rotation of the i-th wheel and motor shaft, and N_v^i is the reduction gear ratio of the angular velocity from the input to output shaft of the i-th actuator.

The relationship between the lateral interaction force F_{yi} of the i-th tire and the load force $F_{\delta i}$ of the piston of the active hydro-cylinder (Fig. 2.30) is established in the following way. It is known that the realization of the tire lateral force F_{yi} is connected with the realization of the rotating moment $\mathcal{M}_{\delta_i}$ about the tire vertical symmetry axis. The existence of this moment is a consequence of the fact that the lateral force F_{yi} acts at a point shifted from the center (see Fig. 2.18) by a distance t_i. As is known, this distance is called the tire pneumatic trail [28]. Hence the relationship between the lateral force and selfaligning torque can be expressed by the relation:

$$F_{yi} = \frac{1}{t_i} \mathcal{M}_{\delta_i} \tag{2.96}$$

Numerical values of the variable t_i for the known values of F_{yi}, $\mathcal{M}_{\delta_i}$, μ_i and α_i can be determined from the Gough diagram (see Fig. 2.20b). By analyzing Fig. 2.30 we can determine functional relation between the tire selfaligning torque $\mathcal{M}_{\delta_i}$ and load force $F_{\delta i}$ of the piston of the hydro-cylinder. This relation can be expressed as:

$$\mathcal{M}_{\delta_i} \cong \frac{1}{1 + \kappa_i} F_{\delta i} l_{\delta i} \tag{2.97}$$

where κ_i is the coefficient of Coulomb's friction at the leverage joints; $l_{\delta i}$ is the geometrical parameter of the mechanism shown in Fig. 2.30. The kinematic relation connecting piston displacement in the hydro-cylinder $e_{\delta i}$ and the value of ground steering angle is given by the approximate expression:

$$e_{\delta i} \cong l_{\delta i}\delta_i \tag{2.98}$$

where δ_i is the ground steering angle of the i-th tire expressed in $[rad]$. By combining the equalities (2.96) and (2.97) we obtain an explicit relation between the lateral tire force and the load force of the piston in the control hydro-cylinder:

$$F_{yi} \cong \frac{1}{t_i}\frac{1}{(1+\kappa_i)}l_{\delta i}F_{\delta i} = \hat{k}_i F_{\delta i} \tag{2.99}$$

The dependence of the payload P_i on the active damping force f_{a_i} is given by the relation (2.42). The corresponding kinematic relations concerning the SS displacement and vertical displacement of the vehicle MC are given by the relations (2.41).

By introducing the relations (2.94) and (2.99) into the equations (2.46)/(2.47) and after additional arranging the vector (2.33), we obtain a relationship between the generalized forces and moments τ and the control variables f_{a_i}, $F_{\delta i}$ and $\mathcal{M}_{\omega i}$ at the output elements of the vehicle actuators. After arranging the expressions we obtain:

$$\tau = J^{-T}(q,\dot{q},d^*)\hat{\tau} + \Gamma(q,\dot{q},d^*,\Delta) \tag{2.100}$$

where q is the (6×1) vector of vehicle state coordinates; J^{-T} is the (6×12) mapping matrix linking the vectors of forces and moments acting at the vehicle body MC and the corresponding vector of forces and moments of the system actuators; Γ is the (6×1) vector of external forces and moments reduced to the vehicle body MC. These forces and moments are consequences of the SS vertical dynamics including also the action of the disturbance Δ produced by the unevenness of the road surface; d^* is the vector of geometrical and dynamic parameters of the mechanical part of the system and parameters of tire-road interaction; vector of control variables $\hat{\tau}$ being defined in the following way:

$$\hat{\tau} = [f_{a_1} \ \cdots \ f_{a_4} \ \mathcal{M}_{\omega 1} \ \cdots \ \mathcal{M}_{\omega 4} \ F_{\delta 1} \ \cdots \ F_{\delta 4}]^T \tag{2.101}$$

In the relation (2.100) the matrix J^{-T} and vector Γ have the following forms:

$$J^{-T}(q,d^*) = \begin{bmatrix} 0 & 0 & j_{3,1} & 1 & j_{5,1} & j_{6,1} \\ 0 & 0 & j_{3,2} & 1 & j_{5,2} & j_{6,2} \\ 0 & 0 & j_{3,3} & 1 & j_{5,3} & j_{6,3} \\ 0 & 0 & j_{3,4} & 1 & j_{5,4} & j_{6,4} \\ j_{1,5} & j_{2,5} & j_{3,5} & 0 & j_{5,5} & j_{6,5} \\ j_{1,6} & j_{2,6} & j_{3,6} & 0 & j_{5,6} & j_{6,6} \\ j_{1,7} & j_{2,7} & j_{3,7} & 0 & j_{5,7} & j_{6,7} \\ j_{1,8} & j_{2,8} & j_{3,8} & 0 & j_{5,8} & j_{6,8} \\ j_{1,9} & j_{2,9} & j_{3,9} & 0 & j_{5,9} & j_{6,9} \\ j_{1,10} & j_{2,10} & j_{3,10} & 0 & j_{5,10} & j_{6,10} \\ j_{1,11} & j_{2,11} & j_{3,11} & 0 & j_{5,11} & j_{6,11} \\ j_{1,12} & j_{2,12} & j_{3,12} & 0 & j_{5,12} & j_{6,12} \end{bmatrix}^T \tag{2.102}$$

and

$$\Gamma(q,\dot{q},d^*,\Delta) = \begin{bmatrix} 0 \\ 0 \\ \Gamma_{3,1} \\ \sum_{i=1}^{4}(P_{ci}+P_{bi}) \\ \Gamma_{5,1} \\ K_6\sum_{i=1}^{2}(P_{ci}+P_{bi})+K_5\sum_{i=3}^{4}(P_{ci}+P_{bi}) \end{bmatrix} \qquad (2.103)$$

$$\begin{aligned}
\Gamma_{3,1} &= (\theta K_1 - \phi K_6)(P_{c1}+P_{b1}) + (\theta K_2 - \phi K_6)(P_{c2}+P_{b2}) + \\
&+ (\theta K_1 - \phi K_5)(P_{c3}+P_{b3}) + (\theta K_2 - \phi K_5)(P_{c4}+P_{b4}) \\
\Gamma_{5,1} &= K_1(P_{c1}+P_{b1}) + K_2(P_{c2}+P_{b2}) + K_1(P_{c3}+P_{b3}) + K_2(P_{c4}+P_{b4})
\end{aligned}$$

Elements of the matrix J^{-T} are calculated from the following expressions:

$$\begin{aligned}
j_{1,5} &= a_1^1; \quad j_{1,6} = a_1^2; \quad j_{1,7} = a_1^3; \quad j_{1,8} = a_1^4; \\
j_{1,9} &= a_2^1; \quad j_{1,10} = a_2^2; \quad j_{1,11} = a_2^3; \quad j_{1,12} = a_2^4; \\
j_{2,5} &= b_1^1; \quad j_{2,6} = b_1^2; \quad j_{2,7} = b_1^3; \quad j_{2,8} = b_1^4; \\
j_{2,9} &= b_2^1; \quad j_{2,10} = b_2^2; \quad j_{2,11} = b_2^3; \quad j_{2,12} = b_2^4; \\
j_{3,1} &= \theta K_1 - \phi K_6; \\
j_{3,2} &= \theta K_2 - \phi K_6; \\
j_{3,3} &= \theta K_1 - \phi K_5; \\
j_{3,4} &= \theta K_2 - \phi K_5; \\
j_{3,5} &= K_9 b_1^1 + K_{11} a_1^1 + \theta K_3 b_1^1 - \phi K_7 a_1^1; \\
j_{3,6} &= K_9 b_1^2 + K_{12} a_1^2 + \theta K_3 b_1^2 - \phi K_7 a_1^2; \\
j_{3,7} &= K_{10} b_1^3 + K_{11} a_1^3 + \theta K_4 b_1^3 - \phi K_8 a_1^3; \\
j_{3,8} &= K_{10} b_1^4 + K_{12} a_1^4 + \theta K_4 b_1^4 - \phi K_8 a_1^4; \\
j_{3,9} &= b_2^1 + a_2^1 + \theta K_3 b_2^1 - \phi K_7 a_2^1; \\
j_{3,10} &= b_2^2 + a_2^2 + \theta K_3 b_2^2 - \phi K_7 a_2^2; \\
j_{3,11} &= b_2^3 + a_2^3 + \theta K_4 b_2^3 - \phi K_8 a_2^3; \\
j_{3,12} &= b_2^4 + a_2^4 + \theta K_4 b_2^4 - \phi K_8 a_2^4; \\
j_{5,1} &= K_1; \quad j_{5,2} = K_2; \quad j_{5,3} = K_1; \quad j_{5,4} = K_2; \\
j_{5,5} &= K_3 b_1^1 - \theta K_9 b_1^1 - \theta K_{11} a_1^1; \\
j_{5,6} &= K_3 b_1^2 - \theta K_9 b_1^2 - \theta K_{12} a_1^2; \\
j_{5,7} &= K_4 b_1^3 - \theta K_{10} b_1^3 - \theta K_{11} a_1^3; \\
j_{5,8} &= K_4 b_1^4 - \theta K_{10} b_1^4 - \theta K_{12} a_1^4; \\
j_{5,9} &= K_3 b_2^1 - \theta b_2^1 - \theta a_2^1; \\
j_{5,10} &= K_3 b_2^2 - \theta b_2^2 - \theta a_2^2; \\
j_{5,11} &= K_4 b_2^3 - \theta b_2^3 - \theta a_2^3; \\
j_{5,12} &= K_4 b_2^4 - \theta b_2^4 - \theta a_2^4;
\end{aligned}$$

$$
\begin{aligned}
j_{6,1} &= K_6; \quad j_{6,2} = K_6; \quad j_{6,3} = K_5; \quad j_{6,4} = K_5; \\
j_{6,5} &= K_7 a_1^1 + \phi K_9 b_1^1 + \phi K_{11} a_1^1; \\
j_{6,6} &= K_7 a_1^2 + \phi K_9 b_1^2 + \phi K_{12} a_1^2; \\
j_{6,7} &= K_8 a_1^3 + \phi K_{10} b_1^3 + \phi K_{11} a_1^3; \\
j_{6,8} &= K_8 a_1^4 + \phi K_{10} b_1^4 + \phi K_{12} a_1^4; \\
j_{6,9} &= K_7 a_2^1 + \phi b_2^1 + \phi a_2^1; \\
j_{6,10} &= K_7 a_2^2 + \phi b_2^2 + \phi a_2^2; \\
j_{6,11} &= K_8 a_2^3 + \phi b_2^3 + \phi a_2^3; \\
j_{6,12} &= K_8 a_2^4 + \phi b_2^4 + \phi a_2^4;
\end{aligned}
\tag{2.104}
$$

where

$$
\begin{aligned}
a_1^i &= \frac{1}{r_i} \cos(\delta_i) \quad i = 1, \ldots, 4 \\
a_2^i &= -\hat{k}_i \sin(\delta_i) \quad i = 1, \ldots, 4 \\
b_1^i &= \frac{1}{r_i} \sin(\delta_i) \quad i = 1, \ldots, 4 \\
b_2^i &= \hat{k}_i \cos(\delta_i) \quad i = 1, \ldots, 4 \\
K_1 &= +\left(\frac{s_b}{2} + h_2 \phi\right) \\
K_2 &= -\left(\frac{s_b}{2} - h_2 \phi\right) \\
K_3 &= h_1 - h_5 \theta + l_1 \theta \\
K_4 &= h_1 - h_5 \theta - l_2 \theta \\
K_5 &= +(l_2 + h_4 \theta) \\
K_6 &= -(l_1 - h_4 \theta) \\
K_7 &= -(h_1 - h_5 \theta + l_1 \theta) \\
K_8 &= -(h_1 - h_5 \theta - l_2 \theta) \\
K_9 &= +(l_1 - h_4 \theta) \\
K_{10} &= -(l_2 + h_4 \theta) \\
K_{11} &= -\left(\frac{s_b}{2} + h_2 \phi\right) \\
K_{12} &= +\left(\frac{s_b}{2} - h_2 \phi\right)
\end{aligned}
\tag{2.105}
$$

Taking into account the relations (2.28) and (2.100) we can finally write the integrated model of vehicle dynamics as:

$$
\begin{aligned}
H(q, d)\ddot{q} + h(q, \dot{q}, d) &= J^{-T}(q, d^*)\hat{\tau} + \mathcal{F}(q, \dot{q}, d^*, \Delta) \\
\mathcal{F}(q, \dot{q}, d^*, \Delta) &= F(q, \dot{q}, d) + \Gamma(q, \dot{q}, d^*, \Delta)
\end{aligned}
\tag{2.106}
$$

The vector $\mathcal{F}$ in the model (2.106) represents a (6×1) vector of external resistance forces/moments by which the environment acts on the vehicle body. The vector $\mathcal{F}$ has the form:

$$
\mathcal{F} = [F_X \quad F_Y \quad M_Z \quad F_Z \quad M_X \quad M_Y]^T
\tag{2.107}
$$

In the relation (2.106), the vector of vehicle control forces and moments $\hat{\tau}$ represents a (12×1) vector of driving forces/moments generated by the vehicle's actuators. Finally, it should be mentioned that the scalar variable F_{o_i} that appears in the actuator model (2.93) can take the elements from the vector (2.101).

APPENDIX

PARAMETERS[20] of ROAD VEHICLE MODEL and TIRE-ROAD INTERACTION

parameter	value	units	meaning
C_0	$9.656 \cdot 10^6$	$[kg/m^3]$	stiffness constant of tire-road interface
r_{1_0}	0.30	$[m]$	original radius of the 1st tire (unloaded)
r_{2_0}	0.30	$[m]$	original radius of the 2nd tire (unloaded)
r_{3_0}	0.30	$[m]$	original radius of the 3rd tire (unloaded)
r_{4_0}	0.30	$[m]$	original radius of the 4th tire (unloaded)
w_t	0.20	$[m]$	width of tire
ν_{s_0}	1.00	/	static friction coefficient
ν_{d_0}	0.70	/	sliding friction coefficient
f_r	0.02	/	rolling resistance coefficient of the tire
M_t	13	$[kg]$	tire mass
B_{t_1}	1916	$[N/(m/s)]$	damping coefficient of the 1st tire deflection velocity
B_{t_2}	1916	$[N/(m/s)]$	damping coefficient of the 2nd tire deflection velocity

---

[20]Values were taken from [9]

parameter	value	units	meaning
B_{t_3}	1916	$[N/(m/s)]$	damping coefficient of the 3rd tire deflection velocity
B_{t_4}	1916	$[N/(m/s)]$	damping coefficient of the 4th tire deflection velocity
K_{t_1}	196 200	$[kg/m]$	stiffness coefficient (spring constant) of the 1st tire pneumatics
K_{t_2}	196 200	$[kg/m]$	stiffness coefficient (spring constant) of the 2nd tire pneumatics
K_{t_3}	196 200	$[kg/m]$	stiffness coefficient (spring constant) of the 3rd tire pneumatics
K_{t_4}	196 200	$[kg/m]$	stiffness coefficient (spring constant) of the 4th tire pneumatics
K_{tt_1}	5000	$[N/m^3]$	spring constant of the 1st tire (nonlinear term)
K_{tt_2}	5000	$[N/m^3]$	spring constant of the 2nd tire (nonlinear term)
K_{tt_3}	5000	$[N/m^3]$	spring constant of the 3rd tire (nonlinear term)
K_{tt_4}	5000	$[N/m^3]$	spring constant of the 4th tire (nonlinear term)
h_1	0.05	$[m]$	vertical distance of vehicle body MC from the centre of a syspension system joint
h_2	0.43	$[m]$	distance from centre of gravity to roll centre
h_4	0.40	$[m]$	z-distance from center of gravity to pitch centre
h_5	0.10	$[m]$	x-distance from center of gravity to pitch centre
l_1	1.15	$[m]$	distance from the center of gravity to front axle
l_2	1.51	$[m]$	distance from the center of gravity to rear axle
z_0	0.60	$[m]$	the original height of the mass centre relative to the ground
s_b	1.50	$[m]$	track
m	1550	$[kg]$	vehicle mass (mass of the structure)
I_x	460	$[kg\ m^2]$	moment of inertia in x-direction
I_y	2300	$[kg\ m^2]$	moment of inertia in y-direction
I_z	3100	$[kg\ m^2]$	moment of inertia in z-direction
K_{s1_1}	40 000	$[N/m]$	spring constant of the 1st tire syspension system (the linear term)

parameter	value	units	meaning
K_{s1_2}	40 000	$[N/m]$	spring constant of the 2nd tire syspension system (the linear term)
K_{s1_3}	40 000	$[N/m]$	spring constant of the 3rd tire syspension system (the linear term)
K_{s1_4}	40 000	$[N/m]$	spring constant of the 4th tire syspension system (the linear term)
K_{s2_1}	40 000	$[1/m^4]$	spring constant of the 1st tire syspension system (the nonlinear term)
K_{s2_2}	40 000	$[1/m^4]$	spring constant of the 2nd tire syspension system (the nonlinear term)
K_{s2_3}	40 000	$[1/m^4]$	spring constant of the 3rd tire syspension system (the nonlinear term)
K_{s2_4}	40 000	$[1/m^4]$	spring constant of the 3rd tire syspension system (the nonlinear term)
B_{s1_1}	10 000	$[N/(m/s)]$	damping coefficient of the viscous damper at the 1st tire SS (the linear term)
B_{s1_2}	10 000	$[N/(m/s)]$	damping coefficient of the viscous damper at the 2nd tire SS (the linear term)
B_{s1_3}	10 000	$[N/(m/s)]$	damping coefficient of the viscous damper at the 3rd tire SS (the linear term)
B_{s1_4}	10 000	$[N/(m/s)]$	damping coefficient of the viscous damper at the 4th tire SS (the linear term)
$\overline{B}_{s2_1}$	4000	$[N/(m/s)]$	damping coefficient of the viscous damper at the 1st tire SS (the nonlinear term)
$\overline{B}_{s2_2}$	4000	$[N/(m/s)]$	damping coefficient of the viscous damper at the 2nd tire SS (the nonlinear term)
$\overline{B}_{s2_3}$	4000	$[N/(m/s)]$	damping coefficient of the viscous damper at the 3rd tire SS (the nonlinear term)
$\overline{B}_{s2_4}$	4000	$[N/(m/s)]$	damping coefficient of the viscous damper at the 4th tire SS (the nonlinear term)
$\overline{w}$	0.025	$[m/s]$	maximal piston's speed in the viscous hydro-cylinder
$T_{d_{hc}}$	0.032	$[s]$	constant of time lag in the hydro-cylinder
$T_{d_{em}}$	0.025	$[s]$	constant of time lag in the electromotor
K_x	0.45	$[Ns^2/m^2]$	air resistance coefficient in the longitudinal direction
K_y	2.10		air resistance coefficient in the lateral direction
K_ε	50 000	$[Nm\,rad/s]$	yaw damping coefficient
g	9.81	$[m/s^2]$	gravity acceleartion
RC	1.00	$/$	characteristic coefficient of tire-road interaction

Bibliography

[1] Satoh, K., Moroizumi, H., Mimuro, T., Tanaka, T., "Modeling Technique and Its Certification for Full Vehicle Handling Simulation", *Proceedings of International Symposium on Advanced Vehicle Control for Active Safety and Ride Comfort AVEC'94*, pp. 79-84, Tsukuba, Japan, 1994

[2] ADAMS Reference Manuel. *Mechanical Dynamics, Inc.*, Ann Arbor, 1991

[3] Körtum, W., Rulka, W., Schwartz, W., "Analysis and Design of Controlled Vehicles Using Multibody Simulation Models", *Proceedings of International Symposium on Advanced Vehicle Control for Active Safety and Ride Comfort AVEC'94*, pp. 85-92, Tsukuba, Japan, 1994

[4] Jansen, S., T., Van Osten, J., M., "The Development and Evaulation of an Active 4WS Vehicle Simulation Model with Limited Complexity Using Driving Tests", *Proceedings of International Symposium on Advanced Vehicle Control for Active Safety and Ride Comfort AVEC'94*, pp. 438-443, Tsukuba, Japan, 1994

[5] Straub, A., "Dynamic Stability Control in BMW 7 Series Cars", *Proceedings of International Symposium on Advanced Vehicle Control AVEC'96*, pp. 547-559, RWTH Aachen, Germany, 1996

[6] Yi, K., Oh, T., Suh, M. W., "A Robust Semi-active Suspension Control to Improve Ride Quality", *Proceedings of International Symposium on Advanced Vehicle Control for Active Safety and Ride Comfort AVEC'94*, pp. 195-199, Tsukuba, Japan, 1994

[7] Irmscher, S., Hees, E., Kutche, T., "A Controlled Suspension System with Continously Adjustable Damping Force", *Proceedings of International Symposium on Advanced Vehicle Control for Active Safety and Ride Comfort AVEC'94*, pp. 325-330, Tsukuba, Japan, 1994

[8] Sekulić, M., R., *Fundamentals of Automatic Control Theory - Servomechanisms, Belgrade: Scientific book*, 1982

[9] Peng, H., Tomizuka, M., "Program on Advanced Technology (PATH) Program - Lateral Control of Front-Wheel-Steering Rubber-Tire Vehicles",

Technical report UCB-ITS-PRR-90-5, Institute of Transportation Studies, University of California at Berkeley, UC Berkeley, 1990

[10] Vallurupalli, S., Dukkipati, R., Osman, M., "Adaptive Active Suspension for a Half-Car Model with a Stochastic Dirt Road Input", *Proceedings of International Symposium on Advanced Vehicle Control for Active Safety and Ride Comfort AVEC'94*, pp. 367-372, Tsukuba, Japan, 1994

[11] Tokuda, H., Murai, T., Nakashima, N., Matsunaga, E., Minabe, H., "Vehicle Body Motion Detection Algorithm Used by Wheel Speed Signal-Application for Suspension Control System", *Proceedings of International Symposium on Advanced Vehicle Control for Active Safety and Ride Comfort AVEC'94*, pp. 539-543, Tsukuba, Japan, 1994

[12] Araki, Y., Oya, M., Harada, H., "Preview Control of Active Suspension Using Disturbance of Front Wheel", *Proceedings of International Symposium on Advanced Vehicle Control for Active Safety and Ride Comfort AVEC'94*, pp. 299-304, Tsukuba, Japan, 1994

[13] Dorling, R., J., Cebon, D., "Achievable Roll Response of Automotive Suspensions", *Proceedings of International Symposium on Advanced Vehicle Control for Active Safety and Ride Comfort AVEC'94*, pp. 409-427, Tsukuba, Japan, 1994

[14] Ackermann, J., Sienel, W., "Robust Yaw Damping of Cars with Front and Rear Wheel Steering", *IEEE Transactions on Control Systems Technology*, Vol. 1, No. 1, pp. 15-20, March 1993

[15] Bolzern, P., Casati, A., Locatelli, A., Maroni, M., "Control of 4WS Vehicles: A Design Based on μ-Synthesis Techniques", *Proceedings of the 13th IFAC Triennal World Congress*, pp. 53-58, San Francisco, USA, 1996

[16] Pham, H., Hedrick, K., Tomizuka, M., "Autonomous Steering and Cruise Control of Automobiles via Sliding Mode Control", *Proceedings of International Symposium on Advanced Vehicle Control for Active Safety and Ride Comfort AVEC'94*, pp. 444-448, Tsukuba, Japan, 1994

[17] Plöhl, M., Lugner, P., "Theoretical Investigations of the Interaction Driver-Feedback-Controlled Automobile", *Proceedings of International Symposium on Advanced Vehicle Control for Active Safety and Ride Comfort AVEC'94*, pp. 42-48, Tsukuba, Japan, 1994

[18] Reikert, P., Schunk, T., "Zur Fahrmechanik des gummibereiften Kraftfahrzeugs", *Ingenieur Archiv*, Vol. 11, pp. 210-224, 1940

[19] Inagaki, S., Kshiro, I., Yamamoto, M., "Analysis of Vehicle Stability in Critical Cornering Using Phase-Plane Method", *Proceedings of International Symposium on Advanced Vehicle Control for Active Safety and Ride Comfort AVEC'94*, pp. 287-292, Tsukuba, Japan, 1994

[20] Shladover, S., E., Desoer, C., A., Hedrick, J., K., Tomizuka, M., Arland, J., Zhang, W., McMahon, D., H., Peng, H., Sheikholeslam, S., McKeown, N., "Automatic Vehicle Control Developments in the PATH Program", *IEEE Transactions on Vehicular Technology*, Vol. 40, No. 1, pp. 114-130, 1991

[21] Hedrick, J., K., Tomizuka, M., Varaiya, P., "Control Issues in Automated Highway Systems", *IEEE Control Systems Magazine*, Vol. 14, No. 6, pp. 21-32, 1994

[22] Lugner, P., "The Influence of the Structure of Automobile Models and Tyre Characteristics on the Theoretical Results of Steady-State and Transient Vehicle Performance", The Dynamics of Vehicles, *Proceedings of the 5th VSD-2nd IUTAM Symposium*, pp. 21-39, 1984.

[23] Kiencke, U., Nielsen, L., Automotive Control Systems For Engine, Driveline, and Vehicle, *Springer-Verlag Berlin Heidelberg New York*,2000

[24] Sakai, H., "Theoretical and Experimental Studies on the Dynamical Properties of Tyres, Part 1: Review of Theories of Rubber Friction", *International Journal of Vehicle Design*, Vol. 2, No. 1, pp. 78-110, 1981

[25] Rodić, A., D., Vukobratović, M., K., "Contribution to the integrated control synthesis of road vehicles", *IEEE Transactions on Control Systems Technology*, Vol. 7, No. 1, pp. 64-78, 1999

[26] Sakai, H., "Theoretical and Experimental Studies on the Dynamical Properties of Tyres, Part 4: Investigations of the Influences of Running Conditions by Calculation and Experiment", *International Journal of Vehicle Design*, Vol. 3, No. 3, pp. 333-375, 1982

[27] Pacejka, H., B., Bakker, E., "The Magic Formula Tyre Model", *Vehicle System Dynamics*, Vol. 21, ISBN 90-265-1332-1, 1993

[28] Pasterkamp, W., R., Pacejka, H., B., "On-line Estimation of Tyre Characteristics for Vehicle Control", *Proceedings of International Symposium on Advanced Vehicle Control for Active Safety and Ride Comfort AVEC'94*, pp. 521-526, Tsukuba, Japan, 1994

[29] Dixon, J., C., Tires, Suspension and Handling, *Warrendale, Pa.: SAE International Inc.*, 1996

[30] Pasterkamp, W., R., Pacejka, H., B., "The Tyre as a Sensor to Estimate Friction", *Proceedings of International Symposium on Advanced Vehicle Control AVEC'96*, pp. 839-854, RWTH Aachen, Germany, 1996

[31] Pacejka, H., B., "Tire modeling for Use in Vehicle Dynamic Studies", *SAE Paper*, Vol. 870421, 1987

[32] A. D. Luca, C. Manes, "modeling of Robots in Contact with the Dynamic Environment", *IEEE Transaction on Robotics and Automation*, Vol. 10, No. 4, pp. 542-548, 1994.

[33] Nicolas, C., F., Landaluze, J., Sabalza, X., Gaston, M., Reyero, R., "An Intelligent Suspension System Based on a Continuously Variable Shock Absorber", *Proceedings of International Symposium on Advanced Vehicle Control AVEC'96*, pp. 241-255, RWTH Aachen, Germany, 1996

[34] Norton, J., P., An Introduction to Identification, *London-Orlando-San Diego-New York-Austin-Montreal-Sydney-Tokyo-Toronto: Academic Press*, 1986

[35] Yoon, S., Han, S., B., "A Study on the Development of Active Damper System with Electric Motor Driven Orifice", *Proceedings of International Symposium on Advanced Vehicle Control for Active Safety and Ride Comfort AVEC'94*, pp. 171-176, Tsukuba, Japan, 1994

[36] Higashiyama, K., Hirai, T., Kakizaki, T., Hiramoto, M., "Development of the Active Damper Suspension", *Proceedings of International Symposium on Advanced Vehicle Control for Active Safety and Ride Comfort AVEC'94*, pp. 331-336, Tsukuba, Japan, 1994

[37] Nametz, J., E., Smith, R., E., Sigman, D., R., "The Design and Testing of a Microprocessor Controlled Four Wheel Steer Concept Car", *SAE Technical paper series*, No. 885087, pp. 1.652-1.658, 1988

[38] Mitschke, M., Ahring, E., "Control Loop for Driver-Vehicle with four Wheel Steering", *Proceedings of International Symposium on Advanced Vehicle Control for Active Safety and Ride Comfort AVEC'94*, pp. 28-35, Tsukuba, Japan, 1994

[39] F. Schlütter, P. Wältermann, "Hierarchical Control Structures for Hybrid Vehicles - Modeling, Simulation, and Optimization", *Preprints of the First IFAC Workshop on Advances in Automotive Control*, pp. 108-114, Monte-Verita, Ascona, Switzerland, March, 1995

[40] Wittmer, Ch., Dietrich, Ph., Guzella, L., "Control Strategies for the ETH Hybrid Vehicle", *Preprints of the First IFAC Workshop on Advances in Automotive Control*, pp. 126-132, Monte-Verita, Ascona, Switzerland, March, 1995

[41] Mayer, T., Schröder, D., "Operating Modes and Control Aspects for a Special Hybrid Drivetrain", *Preprints of the First IFAC Workshop on Advances in Automotive Control*, pp. 120-126, Monte-Verita, Ascona, Switzerland, March, 1995

[42] Merrit, E., H., Hydraulic Control Systems, *New York: John Wiley & Sons*, 1967

[43] Vukobratović, M., Potkonjak, V., Dynamics of Manipulation Robots - Theory and Application, *Springer-Verlag, Berlin, Heidelberg, New York*, 1982

Chapter 3

Control of Automated Road Vehicles

Abstract

A conceptually new control scheme for a fully automated road vehicle based on spatial dynamic vehicle model is described in detail. The proposed controller is based on the concept of centralized dynamic control. The strategy of two-level (tactical and executive) distributed hierarchical control is applied. Two different types of dynamic laws are proposed for control on tactical level: purely positional control and combined position-force control. As for the problem of control on executive (operative) level, two basically different realizations of local controllers (electro-mechanical and electro-hydraulic) are outlined. At the end of the chapter are presented the results of vehicle model simulation using the selected control laws. In the examples, a characteristic trajectory is imposed along which act various external disturbances. Simulation results served as the basis for analyzing the possibilities of the synthesized dynamic controller.

The chapter also presents a hybrid neuro-dynamic control scheme consisting of a dynamic controller synthesized in advance and an additional neuro-compensator based on the application of artificial neural network. It is shown how the back-propagation algorithm can be successfully used to train a multilayered neural network aiming at identification of the unmodeled dynamics of the system considered. Also, it is shown how such a structure of the trained network can be used for the synthesis of the so-called hybrid, neuro-dynamic controller of road vehicle. The obtained simulation results confirmed the benefits of using neural networks in the control tasks. A comparasion is made of the results obtained using pure dynamic control and the hybrid controller with neuro-compensator.

3.1 Concepts of Automatic Vehicle Guidance

Nowadays, we can speak of the existence of two different concepts of automated control of road vehicles: (i) decentralized and (ii) centralized. The decentralized approach to vehicle control is pragmatic in its application, as it reduces to the synthesis of several partial, mutually independent controllers, to control the vehicle dynamics in particular coordinate directions. Hereby, the controllers of longitudinal, lateral and vertical dynamics of the object are synthesized using the simplified planar models described in Chapter 2. Simple summation of control signals from the particular local controllers ensures the desired motion and dynamic behavior of the overall system in all six main directions of motion. Some outstanding results in the synthesis of automated vehicle control based on the decentralized concept of system control have been presented in [1]-[6].

In contrast to decentralized concept of control, in the centralized approach is synthesized an integrated vehicle controller on the basis of integrated, spatial dynamic model of the system. The control signals thus obtained, applying the criterion of uniform distribution, are then distributed onto particular driving-control subsystems of the vehicle, yielding thus its desired motion. This approach has been successfully applied in control of other large-scale dynamic systems such as aircrafts [7] and industrial manipulation robots [8]. The same idea, adapted to the nature of the technical system considered, is used in the present monograph. As a result of this, a centralized dynamic controller intended for a fully automated vehicle guidance was synthesized as an alternative to the existing centralized approach.

3.2 Centralized Control Approach

The highest level of automation with road vehicles represents the synthesis of their integrated active systems. Similar control systems have existed for a long time in airplanes, submarines, various kinds of robots, and other large-scale dynamic systems. A road vehicle represents a complex highly nonlinear multibody dynamic system, consisting of rigid bodies and elastic elements. Such a system possesses a large number of DOFs. Some of these DOFs, the quality of which is essential for the safety and ride comfort, have to be controlled, while some others, like elastic modes of the vehicle chassis or some particular elements of the subsystems, should not be controlled. Vehicle stability, quality of its dynamic behavior and maneuvering capabilities, depend predominantly on the system design and performance of its active control subsystems. The choice of the best control strategy within the limits of technical feasibility represents a complex task, solving of which demands knowledge of the vehicle dynamic behavior under different riding conditions. For these reasons, there appears the necessity of having a most accurate model of the system. However, conventional approach to solving the control problem is to adapt the model complexity to the conditions of applying a selected control procedure. An illustrative example supporting this view is the linear optimal regulator, which solves the typical LQ-problem with respect to the state variables [9] by satisfying the preset performance index. This regulator, as well as similar procedures to minimize the criterion function in the frequency domain or in the space of state coordinates, use linear [9, 10] or bilinear quarter car model [11], described in the previous chapter. A somewhat more complex model is the single-track model, popularly known as "bicycle model" of the vehicle [12, 13]. As explained above, this model approximates the vehicle dynamics by a restricted number of DOFs [14]. It describes the system's dynamics in the longitudinal and lateral directions of motion, the dynamics of the vehicle yawing, as well as rotation of front and rear wheels about their vertical axes. A common characteristic of all the above-mentioned simplified models is that none of them describes in full the overall vehicle dynamics, but only the dynamics in the particular directions of motion. Accordingly, the controllers for the longitudinal and lateral motions are synthesized separately, as well as the controller of the suspension system. The latter serves to minimize vibrations, i.e. undesirable vertical bobbing of the vehicle due to variations of the road surface profile. These independently synthesized controllers are then coupled into one unique control system. The solution that will be proposed in this section is founded on the so-called centralized control approach that represents the synthesis of system controller based on the spatial model of entire vehicle dynamics.

In the text to follow we shall explain how a relatively complex, nonlinear, spatial vehicle model can be implemented to synthesize a dynamic vehicle controller. Such a model is capable of encompassing the nonlinear behavior of the vehicle in a sufficiently broad range of values of its state quantities and input control variables. On the other hand, linearized models are good approximations of the system only in certain, relatively narrow, domains of linear dependence from the state variables. Outside of these domains they are just rough approximations of their real behavior. Besides, the simplified, decoupled models ignore cross-coupling effects between the state variables, so that they have only a limited practical applicability.

By the chosen approach of the model-based control is understood that the real- time estimation of time-variable system parameters should be ensured, as well as the estimation of parameters of the tire-road interactions. The algorithm for estimating the mentioned parameters is based on the known equations of the system model. Which control strategy (i.e. control scheme) will be selected it depends on the control task to be executed. If a road vehicle concerned, several criteria to be satisfied during its motion can be established. These are: (i) motion stability, (ii) ride quality, (iii) ride comfort, (iv) criterion of minimal SS and tire deflection, and (v) criterion of high-level vehicle maneuvering capabilities. By motion stability is meant the stability of the prescribed (desired) vehicle trajectory and the realization of the desired chassis orientation. Ride quality during driving refers to the elimination of sudden, jerky changes of the lateral acceleration of the vehicle and sudden changes of the yaw rate about the vertical vehicle axis. Good quality behavior in that sense means continual, smooth changes of velocities and accelerations of the given state quantities. Ride comfort represents elimination of unpleasant vibrations, i.e. minimization of the vehicle body heave motion with respect to the road surface. Minimal SS deflection and tire deflection relate to the dilations (compression or extension) of the springs of the shock absorbers to dilations of tire pneumatics. Large deformation amplitudes raise the danger of vehicle instability or tire defects due to large dynamic loads. By maneuvering capabilities of a vehicle we mean the technical ability of the vehicle to change the driving course in a rather broad range, quickly and easily, with minimal changes of control quantities. The possibility of satisfying nominal criteria of vehicle control depends on the available vehicle equipment. In that sense we can speak about a fully automated vehicle. Such a vehicle possesses active control systems, sensor systems, and more recently, the system for communication with the "environment". This system ensures information about the road geometry, obstacles, and exact vehicle position on the road. Automated vehicle control can have three hierarchical levels: (i) strategic control level, (ii) tactical control level, and (iii) executive (operative) control level. *Strategic control level*, on the basis of information about the terrain topography and instantaneous position of the vehicle, enables the controller to determine the desired optimal trajectory in the presence of obstacles. On this control level, the use of a vision system, telecommunication, and elements of artificial intelligence is understood. *Tactical control level* determines the "tactics" (way) of realization of the prescribed vehicle trajectory and the dynamics of the relative attitude deflection of the vehicle body w.r.t. road surface during the motion. The executive control level is realized with the aid of the controllers at the level of the system actuators. They produce control forces and torques in the driving/braking subsystems and in the active suspension system. We shall consider here the synthesis of vehicle control only on the tactical and executive control levels.

3.3 Synthesis of Dynamic Vehicle Controller

The dynamic vehicle controller whose synthesis will be described in this section takes simultaneously care of the overall system dynamics in the six main directions of vehicle motion: longitudinal, lateral, vertical, as well as the directions of roll, pitch and yaw angles (see Figs. 2.13 and 2.15). The centralized approach to control, by using the integrated spatial model of the vehicle described in Section 2.3.1 makes the proposed controller suitable for the vehicles with strongly expressed dynamics. Such case appears during cornering at a rather high velocity, or at sudden changes of ride acceleration. It results in higher lateral and yaw accelerations of the vehicle body.

The choice of the appropriate control strategy and the way of its realization represent a delicate problem, solution of which demands sufficiently deep knowledge of the dynamic behavior of the road vehicle under various motion conditions. Thus, a sufficiently faithful vehicle model is needed. The most frequently used models are simplified system models such as, for example, quarter-car vehicle model [9], single-track model [12], planar model of lateral dynamics [19], model of longitudinal dynamics [3, 5], etc. A common characteristic of all these models is that none of them describes the overall vehicle dynamics but only its dynamics in certain directions.

A basic difference between the proposed controller and the already known solutions of automated vehicle guidance systems is in the following. Previous solutions have assumed the synthesis of independent partial controllers in particular directions of vehicle motion and their simple summation into the so-called summated controller, designed without testing the system's stability. Knowing that in certain instances (as for example at a higher forward velocity) a strong coupling of the dynamic effects inside the system between particular DOFs takes place, it can happen that the particular control actions come into collision. Instead of a simple superposition of control signals in the summated controller, with the aim of avoiding the negative mutual influence of control actions in particular directions, we propose the synthesis of the so-called integrated vehicle controller based on the centralized dynamics model (2.28)-(2.54) of the system. By an appropriate choice of control parameters of the integrated vehicle controller, it is possible to guarantee the desired stability of the considered system in the presence of external perturbations and with parameter variations within a predicted range.

The dynamic model (2.28)-(2.54) used in the synthesis of the vehicle controller describes only the most important dynamic effects arising in the system during its motion. Hence, some dynamic effects such as the elastic modes in the vehicle mechanism, dynamics of the vehicle actuators, time delays in the actuators and control subsystems, Coulomb's friction at the SS joints, fluid friction in the hydro-cylinders of the viscous absorbers, etc., were not explicitly modeled. All these phenomena influence the overall dynamic behavior of the road vehicle during its motion. For this reason, in order to control the vehicle while ensuring its full controllability and stability on the road, we proposed a supplementary neuro-compensator of the corresponding system's uncertainties to be added to the existing control structure of the dynamic controller. The compensator architecture based on the artificial multilayer neural network as a sufficiently good identifier of nonlinear functions will be described in the second part of this chapter.

The purpose of the integrated vehicle controller is to ensure the global motion stability of the road vehicle. The vehicle automatic control has also to ensure a satisfactory dynamic behavior of the system, i.e. its satisfactory responses. Sudden changes of the magnitudes of longitudinal and lateral accelerations, i.e. personal feeling of jerky motion, must be avoided. As for the ride comfort, it is very important that the synthesized controller enables sufficient suppression of the vehicle body vibrations due to bobbing motion, as well as the desired minimization of deflections of the SS and tire pneumatics. This means that the controller has a very complex task of satisfying simultaneously different criteria. The control strategy proposed here is the strategy of the so-called distributed hierarchy control. An example of its implementation was described in [15]. The solution which will be presented here is based on the knowledge of the overall vehicle dynamics. Hence, this control strategy may be named as the strategy of centralized dynamic control. The term "hierarchy" is used for the reason that the object control is realized on two levels: tactical and executive. They are separated w.r.t. the existing dynamic modules - the *vehicle body* and *vehicle active suspension system*. The control is "distributed" because the global control signals generated on a higher level are distributed to the lower, executive control level as reference signals. The essence of the proposed

control strategy can be explained as follows. The vehicle controller on the higher
tactical control level, which is based on the information about the errors of global
state variables q and $\dot{q}$ (defined in the model (2.28)), calculates the vector of
generalized forces and torques which have to act at the vehicle MC to realize
the desired motion. The control forces and torques calculated in this way are
realized indirectly by the action of the vehicle actuators operating in the frame of
the executive control level. For a fully automated control of the vehicle motion
and performance, various types of actuators can be applied. They operate in
the scope of the particular vehicle active systems as: active suspension system
(Fig. 2.29), active steering system (Fig. 2.30), and active traction system (Fig.
2.31). The mentioned active systems can be synchronously applied onto all
four wheels, which should give high control performance and enable automatic
control of an autonomous vehicle on the road.

The vehicle autopilot demands exact and timely information about the position of the vehicle on the road, its current velocity, distance from the static
and mobile obstacles along its trajectory, and about the changes of the road geometry parameters. Important prerequisites for the application of the autopilot
system structure to be described below are:

- Nominal vehicle trajectory has to be known in advance, or generated in
  real time during the motion;

- It is possible in every time instant to measure the current position of the
  vehicle body MC sufficiently accurately w.r.t. central line of the road as
  a nominal path. It is also assumed that there are appropriate sensors
  on the vehicle that measure the relative attitude deflections and payload
  magnitudes at the vehicle MC;

- It is possible to measure or observe (calculate) in some way, the time-
  dependent magnitudes of the vehicle body velocities in all six coordinate
  directions of motion;

- The dynamic model presented by the relation (2.28), by its structure and
  complexity, describes sufficiently well the dynamic effects in the system;

- Estimation of the vehicle dynamic parameters and parameters of interaction between the tire pneumatics and road surface is realizable in real
  time. In the case when the above assumptions are valid, the dynamic
  controller on the tactical level can be synthesized.

The dynamic controller of the road vehicle is synthesized on the tactical control level. On this control level are calculated the values of control forces and
torques that have to act at the vehicle body MC (see Fig. 2.13) to realize
a desired motion. The forces/torques calculated according to the criterion of
uniform tireload distribution [16] are distributed as reference values of the actuator forces and moments. These variables are realized by the actuators of
the vehicle active systems, whose control signals are calculated on the executive
level [16]. In view of the errors of the relevant quantities (position-velocity or

force/moment) in the feedback loop of the control scheme, two different types of dynamic control algorithm can be synthesized: pure position control law and combined position-force control law. Both control laws generate the vector of control forces and torques at the vehicle body MC. Which of these two control laws will be implemented in the scope of the vehicle dynamic controller it will depend on the concrete control requirements. Pure positional control is relatively simple for the synthesis because it demands only information about the exact position and velocity of the vehicle body relative to the road surface. On the contrary, in the case of implementation of the combined control law, it is not necessary to measure directly the attitude deflections of the vehicle body. Instead, measurement of the corresponding forces and moments acting at the vehicle body MC w.r.t. its equilibrium state at rest is required.

On the basis of experience gained with road vehicles as large-scale dynamic systems, it can be pointed out that the dynamic interconnections inside the vehicle's mechanism are not of equal intensity in all particular directions. Thus, the longitudinal, lateral, and yaw motions of the vehicle body are mutually strongly coupled. On the other hand, heave motion of the vehicle body during riding is directly dependent on the corresponding displacements in the roll and pitch directions. This is the reason why the considered system (Fig. 2.13) can be dynamically decoupled into two dynamic modules: (i) the vehicle dynamics in the plane of road surface, and (ii) the vehicle dynamics in the conditionally vertical plane. Dynamic interconnections between the DOFs within these dynamic modules are strongly expressed, while the interconnections between the two modules are relatively weak. Having in mind this and taking into account the fact that the system stabilization in longitudinal, lateral and yaw directions (x, y and ε direction, Fig. 2.13) is the most important task of vehicle control, this naturally imposes the necessity of applying the position-velocity control in these motion directions. In the remaining three directions (vertical, roll, and pitch) it is appropriate to apply the force/moment control. In that case, a uniform tireload distribution on the wheels is ensured, with the indirect positive influence on the overall system stability. However, this does not exclude the possibility that the vehicle body motion in all six coordinate directions can be positionally controlled, as we have already demonstrated in [16].

In view of the above the nominal position vector q_0 can be partitioned into two subvectors $q_0^{(1)}$ and $q_0^{(2)}$. Partitioning of the vector of external forces F_0 is carried out in the same way. As mentioned above, the vehicle body possesses $n = 6$ motion DOFs in q directions, so that the vector F of external forces and moments acting upon the vehicle structure is also of order $m = 6$. In the n_1 directions ($n_1 < n$), the vehicle nominal trajectory $q_0^{(1)}$ is prescribed, and the vehicle position and velocity in these directions are directly controlled. At the same time, in the m_2 directions ($m_2 < m$), the variation function of the programmed force/moment $F_0^{(2)}$ is prescribed, so that the forces/moments in these directions are controlled in a direct way. The mentioned vectors $q_0^{(1)}$ and $F_0^{(2)}$ are of dimensions ($n_1 \times 1$) and ($m_2 \times 1$) respectively, and they are prescribed in advance as the programmed (nominal) signals. The remaining two subvectors

$q_0^{(2)}$ and $F_0^{(1)}$ are of dimensions $(n_2 \times 1)$ and $(m_1 \times 1)$, and they are calculated in a direct way using the model (2.73). Having in mind all this, the vectors q_0 and F_0 can be defined in the following partitioned form:

$$q_0 = [q_0^{(1)T} \; q_0^{(2)T}]^T, \quad \text{where} \tag{3.1}$$
$$q_0^{(1)} = [x_0 \; y_0 \; \varepsilon_0]^T \text{ and } q_0^{(2)} = [z_0 \; \phi_0 \; \theta_0 \;]^T$$

$$F_0 = [F_0^{(1)T} \; F_0^{(2)T}]^T, \quad \text{where} \tag{3.2}$$
$$F_0^{(1)} = [F_X^0 \; F_Y^0 \; M_Z^0]^T \text{ and } F_0^{(2)} = [F_Z^0 \; M_X^0 \; M_Y^0 \;]^T$$

Elements of the described vectors belong to the set of real numbers : $q_0^{(1)} \in R^{n_1 \times 1}$, $q_0^{(2)} \in R^{n_2 \times 1}$, $F_0^{(1)} \in R^{m_1 \times 1}$, $F_0^{(2)} \in R^{m_2 \times 1}$. For the considered object of control their dimensions are $n_1 + n_2 = n$, $m_1 + m_2 = m$, $n = m = 6$, $n_1 = n_2 = 3$ and $m_1 = m_2 = 3$.

3.3.1 Tactical Control Level

3.3.1.1. Dynamic Position Control Law

As in the relation (2.29) q denotes a (6×1) vector of global state coordinates (position and orientation of the vehicle body), so $q_0 = [x_0 \; y_0 \; \varepsilon_0 \; z_0 \; \phi_0 \; \theta_0]^T$ represents the vector which denotes the desired (programmed) trajectory of the road vehicle. The trajectory tracking error w.r.t. position and velocity is defined by the relations $\Delta q = q - q_0$ and $\Delta \dot{q} = \dot{q} - \dot{q}_0$, respectively. External and internal disturbances of diverse nature act upon the vehicle, causing deviation of its position from the desired trajectory. The vehicle controller should ensure stability of the system motion along the programmed path, without manual operation action. The vehicle's transition regime in the case of perturbed motion can be of diverse character. It can be influenced by imposing the character of a desired behavior, as will be shown below. The vector of control signals τ can be calculated on the basis of information about the deviations of the state variables Δq and $\Delta \dot{q}$ and on the basis of knowledge of the dynamic vehicle model (2.28). The form of such control law can be expressed in the following way:

$$\tau = \hat{H}(q, d) \left[\ddot{q}_0 + \Pi(\Delta q, \Delta \dot{q}) \right] + \hat{h}(q, \dot{q}, d) - \hat{F}(q, \dot{q}, d) \tag{3.3}$$

$$\Pi(\Delta q, \Delta \dot{q}) = -K_V \, \Delta \dot{q} - K_P \Delta q, \qquad \text{(PD-regulator)}$$

$$\Pi(\Delta q, \Delta \dot{q}) = -K_V \, \Delta \dot{q} - K_P \Delta q - K_I \int_0^t \Delta q \, dt, \quad \text{(PID-regulator)}$$

where $\hat{H}$, $\hat{h}$ and $\hat{F}$ are the corresponding estimated matrices and estimated/measured vectors of the model (2.28); the matrices K_V, K_P and K_I are (6×6) matrices. They represent the coefficient matrices of the vector polynomial Π, defining the character of the families of curves of the desired transient responses. At the same time, they are the matrices of velocity, position, and integral control gains of the PD or PID regulator. The calculated vector τ represents the vector of driving forces and torques which should act at the vehicle MC in order to realize the desired programmed motion q_0 in the longitudinal x, lateral y, and vertical z directions, as well as the stability of the chassis orientation w.r.t. the programmed values of the roll angle ϕ, pitch angle θ, and yaw angle ε about the main axes of inertia.

Values of the gain matrices K_P, K_V and K_I are determined from the characteristic vector equation of the system, so that they satisfy the demands of the system's dynamic behavior along particular coordinate directions in the frequency domain. The procedure for determining the unknown control gains is as follows. If we assume that the state variables can be measured with a desired degree of accuracy and that the dynamic model (2.28) parameters can be determined sufficiently accurately by estimation, then the characteristic equation of the system can be written in a closed-loop form. From this equation, the unknown gains K_P, K_V and K_I can be determined in such way that the system motion $q(t)$ converges towards its nominal, prescribed value $q_0(t)$. In that case the above assumption holds, and it can be taken that $\hat{H}(q) \simeq H(q)$, $\hat{h}(q,\dot{q}) \simeq h(q,\dot{q})$ and $\hat{F}(q,\dot{q}) \simeq F(q,\dot{q})$. By substituting τ from the model equation (2.28) into the control law (3.3) and, introducing the vector polynomial Π in the form of PD or PID controller, two relationships are obtained:

$$\Delta\ddot{q}(t) + K_V \, \Delta\dot{q}(t) + K_P\Delta q(t) = 0, \tag{3.4}$$

$$\Delta\ddot{q}(t) + K_V \, \Delta\dot{q}(t) + K_P\Delta q(t) + K_I \int_0^t \Delta q(t)\, dt = 0$$

By adopting the gain matrices $K_P = diag\{k_P^i\}$, $K_V = diag\{k_V^i\}$ and $K_I = diag\{k_I^i\}$ $(i = 1,\ldots,6)$ in the diagonal form and applying the Laplace operator onto the previous relations, we arrive at the characteristic equation of the closed-loop system, which can take the following two forms:

$$\begin{aligned} f(s) &= s^2 + K_V \, s + K_P, & &\text{for PD-regulator} \\ f(s) &= s^3 + K_V \, s^2 + K_P \, s + K_I & &\text{for PID-regulator} \end{aligned} \tag{3.5}$$

where s is the (6×1) vector of Laplace operands. By finding the "zeros" of the vector characteristic equations (3.5), values of the desired gains K_P, K_V and K_I are determined. The diagonal members of these matrices represent the control gains k_P^i, k_V^i and k_I^i in the specific directions of system motion. By setting various values for these gains one can influence the character of the transient process, as well as the overall dynamic behavior of the system (response rate, overshoot value, etc.). The gains k_P^i, k_V^i and k_I^i are determined in such a way that the angular frequency of the closed-loop system ω_f in the i-th direction lies within the boundaries of its natural frequency $\omega_f \in (0, \omega_n]$, as well as that the system is relatively damped ($\zeta = 0.80 - 0.90$). Also, care should be taken to prescribe a frequency value not close or equal to the resonance system frequency. In the case of a PD controller, in accordance with the previous notes, it can be written that the characteristic equation of the system in the i-th direction is: $f_{PD}^i(s_i) = s_i^2 + 2\,\zeta\,\omega_f\,s_i + \omega_f^2 = 0$. Then the control gain values, in the corresponding coordinate direction i are determined from the relations $k_V^i = 2\,\zeta\,\omega_f$ and $k_P^i = \omega_f^2$. Gains for the PID controller are determined in a similar way as for the PD controller, with the remark that care has to be taken of the frequency ω_f and relative damping coefficient ζ of the dominant pair of poles, while the third pole is placed on the real axis to the left of the preset dominant poles. In that way we performed the synthesis of the vehicle controller on the tactical level, ensuring the desired dynamic behavior of the vehicle. However, the real system trajectory will converge to the programmed one only if the accuracy of dynamic parameters is high enough. Since this is not the case, a robust control has to be synthesized (based on the test of the practical stability), which includes the system's stability under conditions of parameter variation in a certain range, specified in advance. The problem of practical stability of a road vehicle modeled by a "full" dynamic model will be considered in Chapter 4. The control strategy proposed here is represented by the block-scheme given in Fig. 3.1. In the figure, the idea of hierarchical control at two levels is emphasized.

The vehicle should possess three active control systems[1] allowing the realization of vehicle motion in the main six directions. As is evident from Fig. 3.1, the calculated control signal τ is partitioned into two control signals $\tau_I = [0\ 0\ 0\ \tau_4\ \tau_5\ \tau_6]^T$ and $\tau_{II} = [\tau_1\ \tau_2\ \tau_3\ 0\ 0\ 0]^T$. They were obtained by multiplying the vector τ by the diagonal (6×6) selectivity matrices S_1 and S_2. The input, reference signals for the lower executive control level are calculated on the basis of the signals τ_I and τ_{II}. The mentioned reference signals are: active damping force $f_{a_i}^r(t)$ in the i-th viscous cylinder of the ASS, the reference angular velocity of the i-th wheel $\omega_i^r(t)$ and the reference ground steering angle $\delta_i^r(t)$ of the i-th wheel. They have to be realized at the outputs of the vehicle control actuators, in order to ensure the desired motion $(q(t) \to q_0(t))$ in a finite time interval t. The generalized forces τ_I and τ_{II} (see Fig. 3.1) are uniformly distributed onto the four wheels, producing the longitudinal, lateral and vertical force components F_{Ai}, F_{Bi} and P_i $(i = 1, \ldots, 4)$ presented in Fig. 2.13. The mentioned forces should be generated by the vehicle actuators and transmitted through the SS to the vehicle body. It is desirable that the driving torques τ_{II} should be distributed uniformly onto all the four SS joints. This should be done for the reason of more economic energy consumption in the driving actuators, lower dynamic load of the vehicle subsystems, and more uniform wearing of the tire pneumatics. The uniformity criterion of load distribution can be expressed in the form of the following relations:

$$F_{A1}^r = F_{A2}^r = \frac{1}{2}\,\frac{\tau_1}{1 + (l_1/l_2)}$$

$$F_{A3}^r = F_{A4}^r = \frac{1}{2}\,\frac{\tau_1}{1 + (l_1/l_2)}\,\frac{l_1}{l_2}$$

$$F_{B1}^r = F_{B2}^r = \left(\frac{1}{2} - \frac{l_1}{l_1 + l_2}\right)\tau_2 + \frac{1}{l_1 + l_2}\,\tau_3$$

---

[1]These are: active wheel steering (AWS), traction control system (TCS) and active suspension system (ASS).

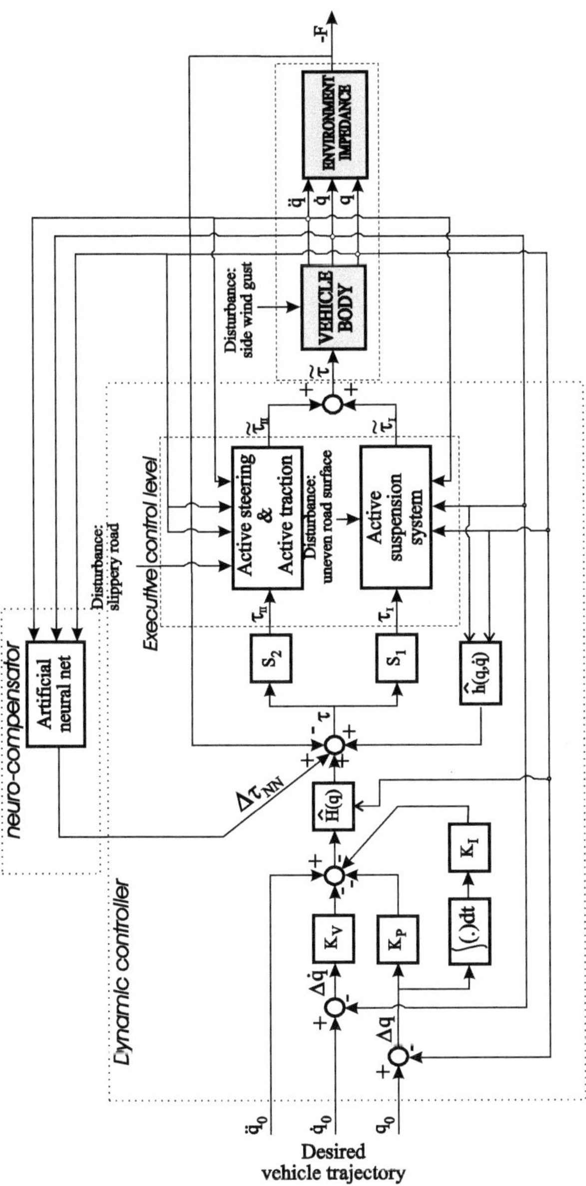

Figure 3.1: Block-scheme of the road vehicle automatic control system with centralized position dynamic controller and supplementary neuro-compensator

$$F_{B3}^r \;\; = \;\; F_{B4}^r = \frac{1}{2} \; \frac{1}{l_1 + l_2} \; (\tau_2 \, l_1 - \tau_3) \tag{3.6}$$

which determine the load distribution onto the individual ASS joints. These forces cause the vehicle motion in the longitudinal and lateral directions. The parameters l_1 and l_2 represent the distances between the vehicle MC and front, i.e. rear wheel axes respectively. The forces $F_{Ai}^r(t)$ and $F_{Bi}^r(t)$ represent the reference force values at the SS joints, on the basis of which the variables ω_i^r and δ_i^r are determined in the further procedure. The variables ω_i and δ_i (for $i = 1, \ldots, 4$) should be changed iteratively in each time sample t within the limits of the realizable values (e.g. between their minimum and maximum values). Based on the knowledge of the nonlinear tire model defined by the relations (2.45)-(2.54), the forces $F_{Ai}(t)$ and $F_{Bi}(t)$ corresponding to the selected pair of variables $\omega_i(t)$ and $\delta_i(t)$, are calculated. Using the method of least squares error the best combination of $\omega_i(t)$, $\delta_i(t)$ $(i = 1, \ldots, 4)$ is detected $(\omega_i(t) \rightarrow \omega_i^r(t)$, $\delta_i(t) \rightarrow \delta_i^r(t))$. It ensures the least sum $(S_i \rightarrow S_i^{min})$ of the squares of errors $S_i = (F_{Ai} - F_{Ai}^r)^2 + (F_{Bi} - F_{Bi}^r)^2$. The obtained values $\omega_i^r(t)$, $\delta_i^r(t)$ represent the reference values of the wheel angular velocity $\omega_i(t)$ and the ground steering angle $\delta_i(t)$ $(i = 1, \ldots, 4)$.

Reference values of the viscous damping active force $f_{a_i}^r$ are determined on the basis of the known value of the vector τ_I using the relations (2.33)-(2.36) and (2.41)-(2.42). The unknown values are calculated by solving the defined system of equations w.r.t. variables f_{a_i} $(i = 1, \ldots, 4)$ in each sampling instant.

The output signals at the lowest hierarchical level represent controlled variables. These outputs of the system actuators $(f_{a_i}(t), \, \omega_i(t), \, \delta_i(t), \, i = 1, \ldots, 4)$ produce the real driving forces and torques $\tilde{\tau}$ (see Fig. 3.1) acting at the vehicle body MC grouped in the vectors $\tilde{\tau}_I$ and $\tilde{\tau}_{II}$. Their vector sum $\tilde{\tau}$ corresponds to the vector of driving forces and torques defined by the relations (2.33)-(2.36). In an ideal case, the vectors τ and $\tilde{\tau}$ are equal, and they differ from each other only to the extent determined by the technical limitations of the actuators, time lag of the actuators, and the inaccuracy of the methods for measuring the vehicle subsystem variables δ, ω, f_a, etc.

Estimation of Model Parameters: In this monograph, we are not going to deal with the procedure of estimating model parameters, as this has been fully covered in the literature. Here we shall only mention those parameters whose estimation is explicitly necessary in the synthesis of dynamic control laws. The determination of these parameters as accurately as possible is an essential prerequisite for good control performance. The application of position control within a dynamic controller of road vehicles requires real-time estimation of the parameters of vehicle model and tire-road interaction. The unknown model parameters for the control synthesis should be determined by on-line estimation during the ride. Apart from the geometrical vehicle parameters, determined by the vehicle design, practically all other parameters are time-variable. It would be quite complicated to estimate all the system parameters, since the estimation is to be performed in real time. Hence, it is indispensable to know which of the vehicle dynamic parameters and parameters of tire-road interaction change more significantly, and what is the range of their changes. First, these are the vehicle mass $\hat{m}$ and moments of inertia about its main axes $\hat{I}_x$, $\hat{I}_y$ and $\hat{I}_z$. The SS parameters and tire pneumatics parameters are also changing with temperature, fluid viscosity in the hydro-cylinders, tire pressure, etc. Changes of these parameters during motion are relatively small provided the vehicle is in good working conditions. Their influence on the dynamic system behavior is practically negligible compared with the effects produced by the change of vehicle mass and inertia. For this reason they can be conditionally taken as being constant. The parameters of the road surface state, i.e. the parameters of tire-road interaction greatly influence the dynamics of vehicle motion, as they may exhibit significant variations. This refers primarily to the sliding friction coefficient $\hat{\nu}_{di}$ ($i = 1,\ldots,4$) and the rolling resistance coefficient $\hat{f}_{ri}$ ($i = 1,\ldots,4$). Also, the side elevation angle of the road surface is of variable magnitude, and can essentially influence the lateral and longitudinal dynamics of the vehicle motion. However, the construction standards of a highway, especially if it is intended for an automated traffic regime, prescribe a zero or a very small value and mildly changing surface elevation angle. In [17] and [3] it has been shown that the information about the forthcoming road geometry, under the condition that this is known, can be sent to the vehicle either if it has been measured or known and guaranteed in advance by the road construction standards.

To estimate the mentioned unknown model parameters ($\hat{m}$, $\hat{I}_x$, $\hat{I}_y$, $\hat{I}_z$, $\hat{\nu}_{di}$, $\hat{f}_{ri}$), we used the recursive least squares estimation method. Hereby, we employed the vehicle dynamic model (2.28)-(2.36) in vectorial form, which can be expanded into six scalar differential equations. Also, in order to estimate the tire-road interaction parameters we used the equations describing the tire model given by the relations (2.46)-(2.54). Small variations of the other, non-estimated parameters, can be treated as internal system disturbances producing no significant effect on the dynamics of the system's motion. Successful estimation of tire parameters can be achieved using the previously described Gough's diagram (Fig. 2.20b) in the way described in [18]. Changes of the other model parameters, whose values vary relatively slightly during motion, can be treated as internal disturbances having no significant influence on the vehicle dynamics.

3.2.1.2. Dynamic Position-Force Control Law

The dynamic position-force control law demands the calculation of the position and velocity error of the vehicle body MC w.r.t. their desired (nominal) values, as well as the deviations of external forces and moments acting at the same point. In case of the application of this control law it is not necessary to measure the state variables $q(t)$, $\dot{q}(t)$ in all coordinate directions but only in x, y and ε directions (Fig. 2.13). Similarly, the component of the force-moment vector $F(t)$ has to be measured in the directions z, ϕ and θ. The basic idea of the combined position-force control law is the choice of position-velocity control in certain selected directions, while in the other directions the force-moment control is applied. Practical benefits of implementing this control algorithm in the designed vehicle controller are viewed in the fact that in some directions is easier to measure force and moment than their positions and velocities. Selection of the directions is done according to the criteria of minimal dynamic coupling between the dynamic modules. Thus, the vectors q and F from the model (2.28) can be formally partitioned in the two subvectors:

$$\begin{aligned}
q &= [q^{(1)T}\ q^{(2)T}]^T, \quad F = [F^{(1)T}\ F^{(2)T}]^T, \\
q^{(1)} &= [x\ y\ \varepsilon]^T, \quad q^{(2)} = [z\ \phi\ \theta]^T, \\
F^{(1)} &= [F_X\ F_Y\ M_Z]^T, \quad F^{(2)} = [F_Z\ M_X\ M_Y]^T
\end{aligned} \quad (3.7)$$

in the way it was done with their nominal forms in (3.1) and (3.2). Such partitioning of the vectors q and F is done for the reason that the influence of forces/moments upon the corresponding displacements of the vehicle body ($F^{(1)}$ to the $q^{(2)}$ and $F^{(2)}$ to the $q^{(1)}$) is weak, and because the system's dynamic behavior in these directions can be decoupled. In fact, it means that the position will be controlled in the directions x, y and ε (longitudinal, lateral, and yaw), while in the directions z, ϕ and θ (vertical, roll, and pitch) force-moment control is to be applied.

Having in mind the relation (2.28), the dynamic position-force control law can be given in the following form:

$$\tau = \hat{H}(q,d)\ddot{q}_c + \hat{h}(q,\dot{q},d) - F \qquad (3.8)$$

$$\ddot{q}_c = \left[\begin{array}{c} \ddot{q}_c^{(1)} \\ \ddot{q}_c^{(2)} \end{array} \right]$$

$$\ddot{q}_c^{(1)} = \ddot{q}_0^{(1)} + \Pi(\Delta q^{(1)}, \Delta \dot{q}^{(1)})$$

$$\ddot{q}_c^{(2)} = -\hat{M}_{22}^{-1}[(F_0^{(2)} + \int_0^t Q(\Delta F^{(2)})\ dt) +$$

$$+ \hat{M}_{21}(\ddot{q}_0^{(1)} + \Pi(\Delta q^{(1)}, \Delta \dot{q}^{(1)})) + \hat{L}^{(2)}$$

$$Q(\Delta F^{(2)}) = -K_F^{(2)}(F^{(2)} - F_0^{(2)}) - K_{FI}^{(2)} \int_0^t (F^{(2)} - F_0^{(2)})dt, \quad \text{(PI-reg.)}$$

$$Q(\Delta F^{(2)}) = -K_F^{(2)}(F^{(2)} - F_0^{(2)}), \quad \text{(P-reg.)}$$

$$\Pi(\Delta q^{(1)}, \Delta \dot{q}^{(1)}) = -K_V^{(1)}(\dot{q}^{(1)} - \dot{q}_0^{(1)}) - K_P^{(1)}(q^{(1)} - q_0^{(1)}), \quad \text{(PD-reg.)}$$

where F is the (6×1) vector of external forces and moments acting at the vehicle body MC; $\hat{M}_{22}$, $\hat{M}_{21}$ and $\hat{L}^{(2)}$ are the estimated matrices of the model (2.73); $Q(.)$ is one of the two disposable (3×1) vector functions which determine the character of the function describing the forces/moments in the transient process; $K_F^{(2)}$ and $K_{FI}^{(2)}$ are the (3×3) matrices of the force control gains of the PI or P-regulator; $F_0^{(2)}$ is the (3×1) vector of the nominal (programmed) values of forces/moments acting in the considered directions (vertical, roll, and pitch); $K_P^{(1)}$ and $K_V^{(1)}$ are quadratic matrices of the position and velocity control gains of dimensions (3×3). These gains act in the directions which are complementary to the directions in which the force/moment control is applied. The feedback control gains included in the dynamic control algorithm (3.8) are determined using a software procedure established on the basis of the practical stability test described in Chapter 4.

The block-scheme of the vehicle autopilot controller is presented in Fig. 3.2. Having in mind its practical implementation, it is assumed that the vehicle has three active control systems ensuring realization of the desired motion in each of the six coordinate directions. For this purpose, three actuators at each wheel of the road vehicle are needed: (i) active viscous cylinder to produce active damping forces in the "vertical" direction (Fig. 2.29); (ii) electric motor[2] which produces rotational moment on the wheel's driveshaft (Fig. 2.31); (iii) viscous hydro-cylinder which by its piston force activates the leverage mechanism to produce tire steering about its vertical axis (Fig. 2.30). Theoretically, with such a choice of system actuators, it is possible to achieve any desired motion. The control vector τ (Fig. 3.2), partitioned in the two complementary vectors of the control signals τ_I and τ_{II}, compensates for the inaccuracies of motion in the corresponding directions so that τ_I acts in the directions z, ϕ and θ, while τ_{II} acts in the x, y and ε directions. The matrices S_1 and S_2 (Fig. 3.2) represent the diagonal (6×6) selectivity matrices serving to separate the control directions. The control signals τ_I and τ_{II} are the reference signals at the lower control level. Control on the executive level depends on the choice of the system's actuators and concrete solution of the active system design. Output variables of the executive control level represent the corresponding signals at the output device of the system actuators. The vehicle actuators generate the driving forces and moments which can be reduced at the vehicle body MC to the vector $\tilde{\tau}$ (Fig. 3.2). In an ideal case, the vectors τ and $\tilde{\tau}$ are identical. But, in reality these vectors differ because of technical limitations of the end-effectors, time delays of the actuators, etc.

Feedback gains in the position-force dynamic control law (3.8) are synthesized in a similar way as the control gains of a pure position controller. If we adopt that the model is ideally known, i.e. that all of its parameters have been accurately determined, then the corresponding gains $K_P^{(1)}$ and $K_V^{(1)}$ in longitudinal, lateral, and yaw directions can be determined from the characteristic equations (3.5) in the previously described way. It remains to determine the force gains $K_F^{(2)}$ in the vertical direction and in the vehicle roll and pitch directions. In a way similar to that described for the case of implementation of pure position control law, the characteristic equation of the system in the three considered directions is obtained by using the Laplace operator on the differential equation of the form $F^{(2)}(t) = F_0^{(2)}(t) - K_F \int_0^t (F^{(2)}(t) - F_0^{(2)}(t))dt$. The obtained characteristic equation is of the form:

$$f(s) = s + K_F^{(2)} = 0 \qquad (3.9)$$

where $K_F^{(2)} = diag\{k_{F_i}^{(2)}\}$ is an $(m_2 \times m_2)$ quadratic diagonal matrix of the

---

[2]Recently, there has been mentioning of "hybrid drivetrain" of road vehicles as a traction system based on combination of an inner combustion (I.C.) motor and an electric motor.

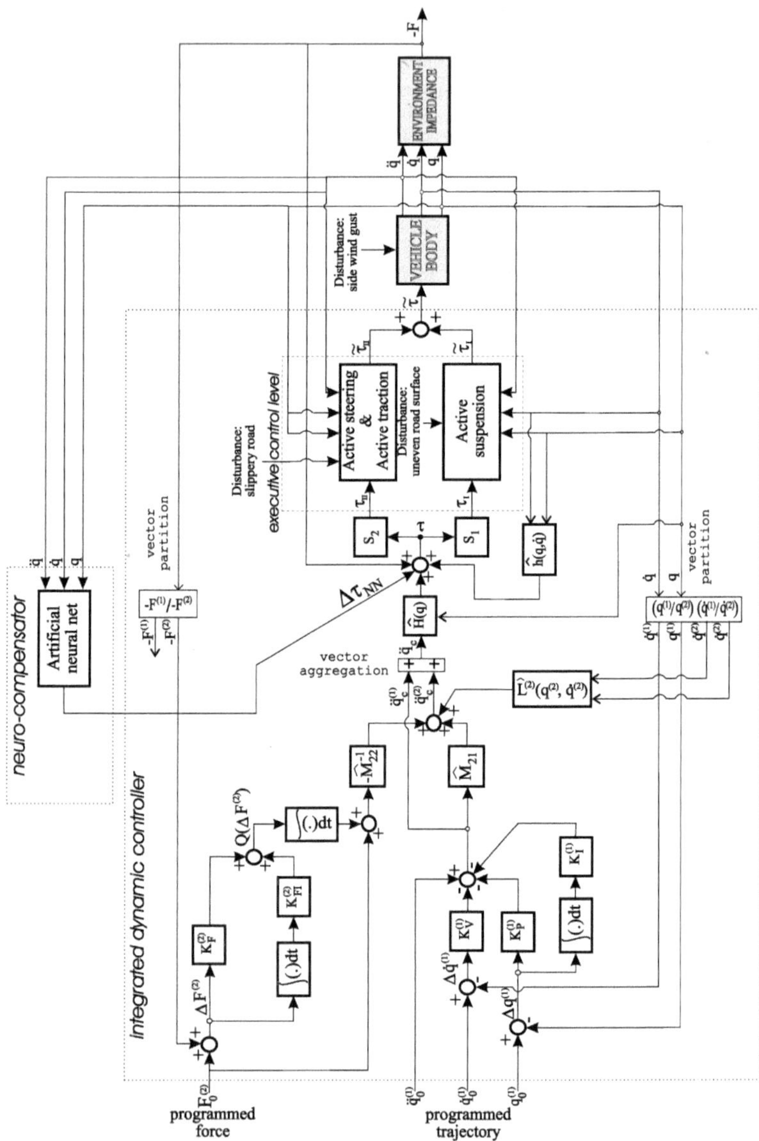

Figure 3.2: Block-scheme of the road vehicle automatic control system with centralized position-force dynamic controller and supplementary neuro-compensator

order $m_2 = 3$. The gains $k_{F_i}^{(2)}$, $i = 1, 2, 3$ are determined so that the closed-loop system poles are given to be at the real, negative part of the abscissa axis in the s-plane, so that the system has the desired dynamic behavior and degree of stability.

3.3.2 Executive Control Level

This control level generates control signals at the level of vehicle actuators so that they realize the forces and torques at their outputs. These actions are then transmitted to the vehicle body, to produce the corresponding forces and torques at the vehicle body MC. In this way the actuators influence indirectly the system motion. The ASS produces an active damping force f_{a_i} (Fig. 2.13) in the i-th hydro-cylinder of the viscous damper, so that the character of the system behavior in the vertical direction is changed. The action of the appropriately synthesized positional controllers of the ASS (Fig. 2.29) distributed onto all four wheels minimizes the amplitudes of the vertical accelerations P_i at the SS joints, which essentially improves the ride comfort. They also diminish the amplitudes of the SS deflections $e_i = z_s^i - z_t^i$ and tire pneumatics deflections $e_{ti} = z_t^i - z_r^i$ (Fig. 2.29). In this way, the unfavorable effects on the wheels and SS joints are lowered, and their service life prolonged.

Lateral motion stability, yawing stability about the vertical axis, as well as the longitudinal motion stability are under direct influence of the distributed control of the driving/braking actuators and the changes of tire-ground steering angles. The generalized forces τ calculated at the hierarchically higher control level are realized by means of changing the longitudinal F_{x_i} and lateral F_{y_i} tire-road interaction forces. By their intensity and direction, these forces influence the changes of the riding direction and vehicle speed, aimed at achieving the desired programmed motion $q_0(t)$. Since the forces F_{x_i} and F_{y_i} are functions of the tire rotation velocity ω_i and the tire ground steering angles δ_i ($i = 1, \ldots, 4$), the control problem reduces to determining the control vectors $\delta(t)$ and $\omega(t)$ in every sampling instant t.

The calculated variables $f_a(t)$, $\omega(t)$ and $\delta(t)$ are realized by using hydro-mechanical or electro-mechanical actuators of different type. Hence, the real control quantities of the system are not those mentioned above. Conventionally, these are input solenoid current of a hydro-cylinder servovalve and input voltage of the inductor of an electric motor. Which kind of servo-actuator will be applied it depends on the concrete vehicle construction and design of its subsystems. The proposed control scheme of the vehicle autopilot can be realized by selecting the following actuators. A four-wheel driving (4WD) vehicle can be designed as a serial hybrid drivetrain [19] with an IC engine and four supplementary electric motors powering all four wheels. Independent four-wheel steering (I4WS) can be realized by four hydro-cylinders (Fig. 2.30) that rotate four mounting brackets resulting in steering all the four car wheels. A similar solution is used with Ford 4WS vehicle with power steering and automatic transmission [20].

To control a hydro-cylinder means to realize the desired force on its piston, which will exceed the external load. The cylinders used in the active suspension system and active steering system are double-acting and represent control objects with typical aperiodic system response. Hence, their dynamic model is nonlinear [21]. However, the dynamic object characteristics can be fairly well approximated by the transfer function [22]:

$$G_{ob}(s) = \frac{K}{T_{ob}s + 1} e^{-\tau_{ob}\,s} \tag{3.10}$$

where K represents the object static gain, T_{ob} is the time constant, and τ_{ob} is the transport object delay in the hydro-cylinder. For a more exact approximation of the object dynamic characteristics with aperiodic step response it is possible to use the procedure illustrated in Fig. 3.3a. Let us assume that the values of

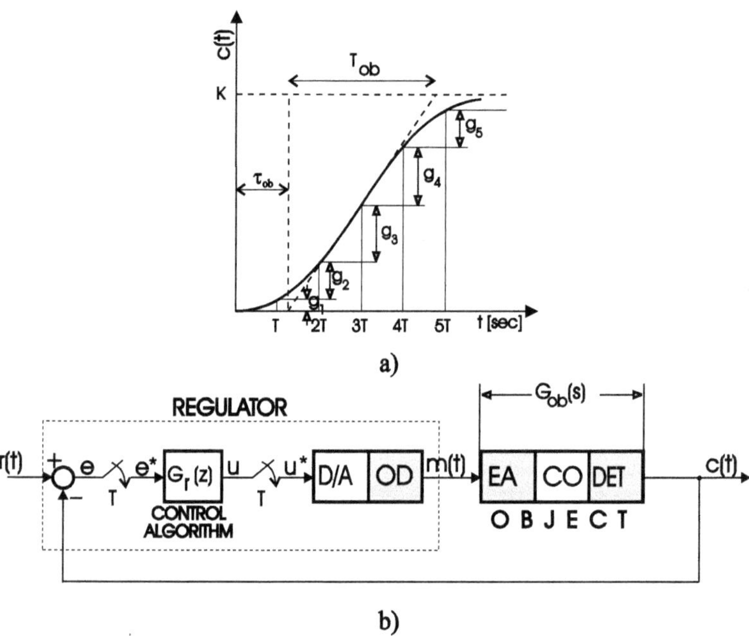

Figure 3.3: a) Form of the unit step response of the hydro-cylinder used in the active subsystems of a road vehicle, b) Control system structure with PID regulator for controlling a hydro-cylinder

g_i $(i = 1, 2, \ldots, n)$ are measured at the object step response. Then the object discrete transfer function can be approximated by [23]:

$$G_{ob}(z^{-1}) = g_1(z^{-1}) + g_2(z^{-2}) + \ldots + g_{n-1}(z^{-(n-1)}) + \frac{g_n z^{-n}}{1 - p\,z^{-1}} \tag{3.11}$$

or by

$$G_{ob}(z^{-1}) = \frac{b_1(z^{-1}) + b_2(z^{-2}) + \ldots + b_n(z^{-n})}{1 - p\,z^{-1}}$$

$$p = 1 - \frac{g_n}{K - \sum_{i=1}^{n-1} g_i} \qquad (3.12)$$

where p is a positive constant smaller than 1, called the attenuating factor.

Since the object discrete transfer function $G_{ob}(z^{-1})$ has been determined, the digital PID regulator illustrated in Fig. 3.3b can be synthesized. It is assumed that the digital regulator possesses at its input part an A/D converter generating the error samples $e^* = e(kT)$ of the controlled variable $c(t)$. The regulator possesses a microprocessor $G_r(z)$ which, on the basis of the signals of error samples $e(kT)$ $(k = 1, 2, \ldots)$, realizes the control algorithm described by the transfer function of the positional PID regulator [24] $G_r(z) = K_p + \frac{K_i}{1 - z^{-1}} + K_d(1 - z^{-1})$, where K_p, K_i and K_d are the corresponding factors of the PID actions. The synthesized PID regulator possesses at its output a D/A converter and an output device (OD) which transforms the signal at the D/A converter output into a desired signal, demanded by the end-effector actuator (EA). The EA, control object (CO), and the detector of the controlled variable (DET), taken all together represent the control object $(G_{ob}(s))$ in a broader sense. Tuning of the control gains K_p, K_i and K_d of the digital PID regulator can be done using some of the well-known procedures [24].

The electric drives of the vehicle wheels can be suitably controlled by means of a microprocessor. The control manner itself depends on the electric motor type. Here we are not concerned with the choice of an optimal actuator type as we want only to demonstrate the procedure how the control signals from the higher hierarchical level are used for control on the lower, executive control level. The problem here is treated only on the level of the applyed model, to demonstrate the corresponding effects by simulation. Of course, in real implementations more care must be taken concerning this problem. Since in the above discussion of the tactical control level we explained how the reference values of the wheel angular velocities ω_i^r $(i = 1, \ldots, 4)$ are calculated, it remains to show how in the case of a DC motor[3] we can control its angular velocity $\omega_i(t)$. In Fig. 3.4 is presented the structure of a microprocessor system [25] to control the angular velocity of a DC motor. The system contains a continual control object (electric motor) and a digital part realizing the digital control law.

In the former part, there are the D/A converter, power amplifier (PA) with the output voltage saturation $\pm U$, and the DC motor. The transfer function $G_m(s)$ for the DC motor voltage $U(s)$ across the ends of the rotor winding to the angular velocity $\Omega(s)$ of the motor shaft, can be approximated by a linear model. The D/A converter can be treated as a zero sampling hold of the linear

---

[3]For the vehicles realizing higher driving torques on their wheels it is convenient to use asynchronous motors, which can be of different design and thus have diverse control possibilities.

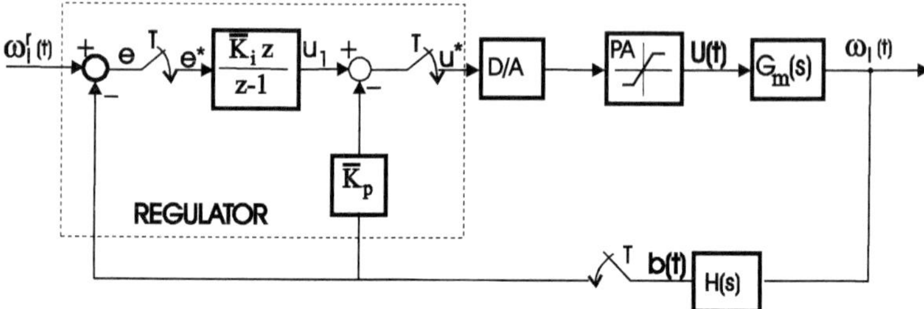

Figure 3.4: Microprocessor control system for a DC motor speed control

function $G_h(s)$:

$$G_m(s) = \frac{\Omega(s)}{U(s)} = \frac{K_m}{T_m s + 1}; \qquad G_h(s) = K_h \frac{1 - e^{-Ts}}{s} \tag{3.13}$$

where K_m and K_h are the corresponding gain factors, T_m is the mechanical time constant, and T is the conversion time. The transfer function of the feedback elements $H(s)$, taking into account the behavior of the type of the counting incremental encoder, is:

$$H(s) = \frac{B(s)}{\Omega(s)} = \frac{K_n}{2\pi} \frac{1 - e^{-Ts}}{s} \tag{3.14}$$

where K_n is the number of markers on the encoder disk. In Fig. 3.4 is presented the control structure with a serial compensator of digital integrator type involving proportional action in the local feedback loop. By shifting the P-action into the feedback loop, abrupt jumps of the control variable $u(kT)$ are avoided in the instants when the preset value of the control variable $\omega_i(t)$ changes stepwise. The proposed control law can be written in the form of the three relations suitable for the programmed realization using a microprocessor:

$$
\begin{aligned}
e(kT) &= \omega_i^r(kT) - b(kT), \\
u_1(kT) &= \overline{K}_i e(kT) + u_1(kT - T), \\
u(kT) &= -\overline{K}_p b(kT) + u_1(kT)
\end{aligned}
\tag{3.15}
$$

where $e(kT)$ and $u_1(kT)$, respectively, are the error samples and output signals of the integral compensator, while $b(kT)$ is the sample of the incremental encoder output in the k-th sampling instant. When the PA is working in a linear regime with the gain K_A, the characteristic equation of the system can be derived relatively easily [25] on the basis of the block-diagram given in Fig. 3.4. It is of the form:

$$
\begin{aligned}
f(z) &= z^3 + a_2 z^2 + a_1 z + a_0 = 0 \\
a_2 &= -(1+A) + \overline{K}(T - T_m + A T_m)(\overline{K}_p + \overline{K}_i) \\
a_1 &= A - \overline{K} T A(\overline{K}_p + \overline{K}_i) - \overline{K} T \overline{K}_p + \overline{K} T_m (1 - A)(2\overline{K}_p + \overline{K}_i) \\
a_0 &= \overline{K} T A \overline{K}_p - \overline{K} T_m \overline{K}_p (1 - A) \\
\overline{K} &= K_h K_A K_m (K_n / 2\pi) \\
A &= e^{-T/T_m}
\end{aligned}
\tag{3.16}
$$

These coefficients depend on the fixed system parameters and the gains of the P- and I- actions, which can be tuned in such a way that the system acquires the desired dynamic characteristics.

3.3.3 Simulation Experiments with Proposed Controller

Performance of the closed-loop control system is analyzed on the basis of simulation of the synthesized dynamic vehicle controller. For the purpose of simulation use was made of the spatial, nonlinear system model whose behavior is described by the relations (2.28)-(2.54), and whose geometrical and dynamic parameters are given in Appendix of Chapter 2. In the simulation experiment the purely positional dynamic controller (3.3) of vehicle motion was used, represented by the control scheme shown in Fig. 3.1. The combined position-force control law will be simulated in the example in Chapter 4.

3.3.3.1 Setting the Simulation Conditions

We simulated the road vehicle model under the condition of a characteristic maneuver in which the longitudinal and lateral object dynamics are especially pronounced. The aim of the control applied is to stabilize tracking of the given trajectory (both in respect of position and force) and to ensure the high quality of dynamic behavior of the system in main six directions of motion. The nominal curvilinear trajectory used in simulation is presented in Fig. 3.5. The vehicle has to track the desired path at a given constant forward velocity $V = 80$ [km/h]. In its motion along the given path (Fig. 3.5a) the vehicle realizes the corresponding accelerations whose profiles are presented in Figs 3.5b and 3.5c. The form of the desired trajectory with the vehicle's velocities in particular coordinate directions is given in Fig. 3.6. In the motion thus defined it was requested that the vehicle retain an unchanged relative position with respect to the road surface plane. Also, it was demanded that, irrespective of the changes in the road surface profile, the vehicle preserves unchanged its relative position $z_0(t) = 0.60$ [m] with respect to the "ideal plane" of road surface, and that the angles of the vehicle body roll and pitch are zero $\phi_0(t) = \theta_0(t) = 0$ [rad]. This means that the corresponding velocities and accelerations in the mentioned three coordinate directions have zero values.

The influence of the external disturbances acting on the vehicle during motion was simulated. The disturbances considered were: variations in the road

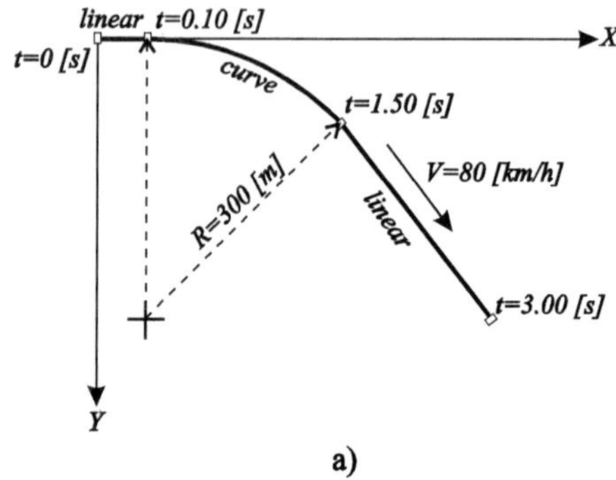

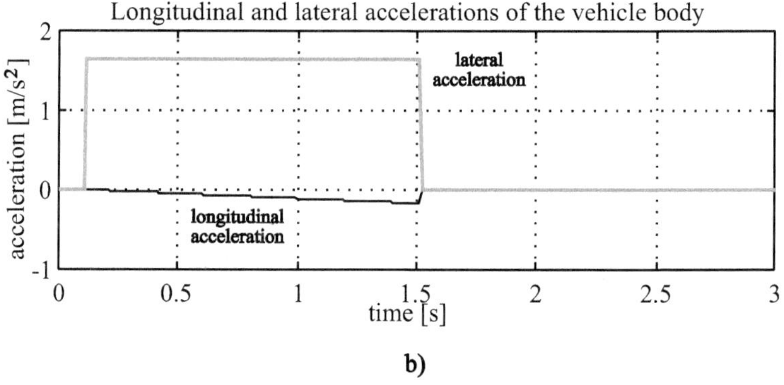

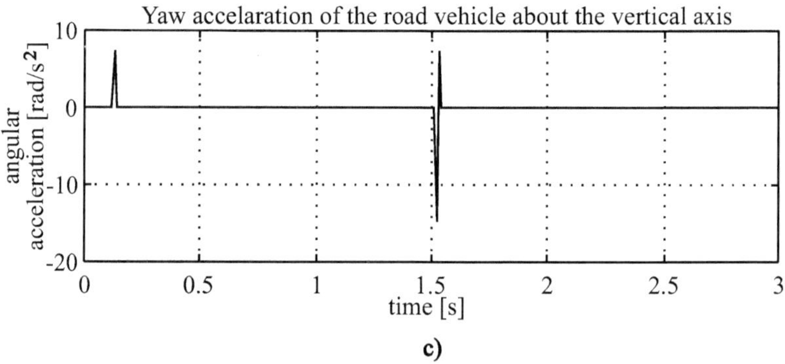

Figure 3.5: a) Geometrical parameters of the nominal vehicle path with the given times of passing through the characteristic points, b) Nominal vehicle's acceleration in longitudinal and lateral directions c) Acceleration of the vehicle yawing about its vertical Z-axis

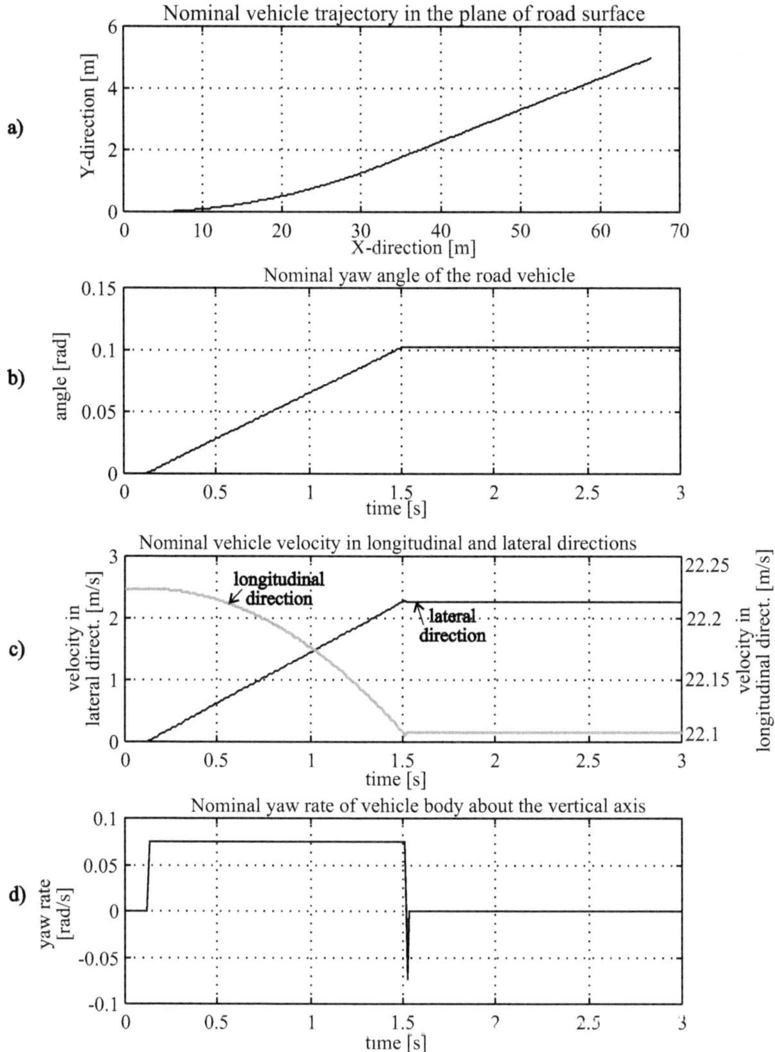

Figure 3.6: a) Vehicle nominal trajectory in the road surface plane, b) View of the nominal yaw angle, c) Nominal velocities in the longitudinal and lateral directions of motion, d) Nominal angular velocity of the vehicle yawing

surface profile $z_r(t)$ (Fig. 3.7b), variations of the road surface side elevation angle $\gamma(t)$, sudden change of the tire sliding coefficients $\nu_d(t)$ of the wheels running on a slippery surface, and a sudden side wind gust (Fig. 3.7a) of variable intensity $F_w(t)$.

The model of road surface profile was given as a random function with a maximum amplitude $z_r^{max} = 1$ $[cm]$ and frequency $f_{z_r} = 22$ $[Hz]$. Hereby, it was assumed that the front and rear wheel pairs follow the same path and that the same perturbation signals appear on them. At that, there is a time difference by the amount of the time constant, determined on the basis of the vehicle motion speed V. In the model shown in Figs. 2.13 and 2.15, the leading wheels are denoted as "1" and "2", whereas the tracking ones are denoted by "3" and "4". A small sinusoidal variation of side elevation angle of the road surface was also given as $\gamma(t) = 2 \sin \omega_\gamma t$ $[deg]$. Here, ω_γ is the angular frequency of the angle amplitude variation calculated from the condition that the side elevation angle is changing periodically at each 2 $[km]$. Changes of the vehicle dynamics parameters m, I_x, I_y and I_z of ± 25 % during the motion were allowed, as a consequence of fuel consumption and load rearrangement due to passengers dislocation in the vehicle. Estimation of the unknown parameters $\hat{m}$, $\hat{I}_x$, $\hat{I}_y$ and $\hat{I}_z$ was carried out during simulation. The efficiency of the controller presented by the control scheme shown in Fig. 3.1 was analyzed on the basis of the simulation results obtained for three characteristic cases: (i) appearance of a variable sliding coefficient, (ii) occurrence of a sudden side wind gust, and (iii) case of vehicle cornering with 25 % higher speed along the same trajectory. All three cases were simulated under the same conditions of motion along a prescribed trajectory. Hereby, it was assumed that the vehicle deviates from the prescribed trajectory in the initial instant, e.g. it is lagging in the longitudinal direction by $\Delta x = -0.20$ $[m]$, has a lateral position error of $\Delta y = -0.10$ $[m]$, and has an error of the relative vertical position of the vehicle MC w.r.t the road surface $\Delta z = 5$ $[mm]$. The errors of initial orientation angles w.r.t. their equilibrium positions were also imposed to be: $\Delta \phi = +0.01$ $[rad]$, pitch $\Delta \theta = -0.01$ $[rad]$, and yaw $\Delta \varepsilon = -0.03$ $[rad]$. In case (i) it was prescribed that during cornering the vehicle encounters a slippery road, whereby the dynamic sliding friction coefficient reduces radically from $\nu_d = 0.7$ to $\nu_d = 0.21$ in a time interval of motion from 0.5 to 1.0 $[s]$ from the initial moment simulated. The new value of the unknown sliding coefficient $\hat{\nu}_d$ was determined by on-line estimation in the way that was outlined in the previous section. In case (ii) the influence of a sudden side wind gust of constant direction ($\psi_{wind} = 60$ $[deg]$), clockwise w.r.t. the X-direction of motion was simulated. Hereby, it was prescribed that the wind intensity decreases according to the law $F_w(t) = 5000 \, exp^{-\lambda \Delta t}$ $[N]$ in the time interval $\Delta t \in (0.3, 1.3]$ $[sec]$. The wind abating parameter λ was calculated from the condition that the intensity of wind gust reduces to $F_{wind}(t) = 0.1$ $[N]$ in Δt $[sec]$. Simulation involved the vehicle motion on a dry road (with a constant value $\nu_d = 0.7$) and without wind gust (case (iii)). Hereby, the vehicle motion was prescribed along the same trajectory at a constant speed of $V = 100$ $[km/h]$ which is by 25 % higher than the previous nominal value. The idea was to demonstrate by the proposed controller can ensure the same vehicle

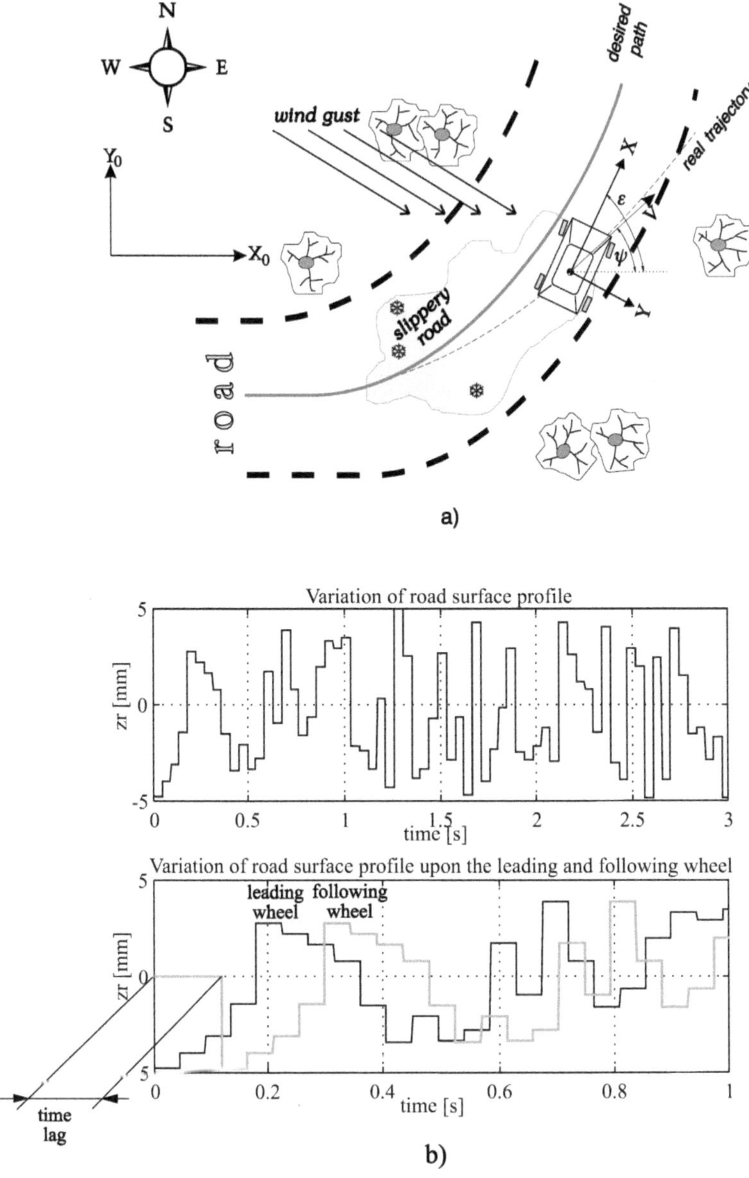

Figure 3.7: The most frequent external disturbances acting upon the vehicle during its motion: a) slippery road and side wind gust, b) variation of road surface profile

trajectory is "mastered" at a higher speed without losing motion stability or ride comfort.

The controller based on the relation (3.3) was synthesized while selecting the gains in such a way that the control system remains stable and possesses the desired dynamic behavior. The gain matrices K_P and K_V have been selected as quadratic, diagonal (6×6) matrices, with the diagonal elements: $k_P^1 = k_P^2 = k_P^3 = 9.8696$, $k_P^4 = k_P^5 = k_P^6 = 355.3058$ $[s^{-2}]$ and $k_V^1 = k_V^2 = k_V^3 = 6.2832$, $k_V^4 = k_V^5 = k_V^6 = 37.6991$ $[s^{-1}]$.

3.3.3.2 Simulation Results

It is well known that even a skilled driver cannot control in explicitly successful way the system behavior in all directions (translatory-angular) simultaneously, which an automatic system is capable of in any case. For this reason, we give an illustration that undoubtedly points to the advantage of the automatic control system to control those coordinates along which the driver is not capable of improving the dynamic behavior of the vehicle under specific driving conditions.

The results of model simulation for the three previously mentioned cases are presented graphically in Figs. 3.8a, 3.8b, 3.9, and 3.10. In Fig. 3.8a is illustrated the quality of the realized motion velocity along the desired trajectory of the vehicle subjected to perturbation of various nature and intensity. The efficiency of the proposed control scheme (Fig. 3.1) was analyzed on the basis of the simulation results (Figs 3.8a, 3.8b) obtained for the case of vehicle motion on a dry road, slippery road, dry road with a side wind gust, and on the dry road with 25 % higher forward velocity compared with the nominal one. As can be seen from Fig. 3.8a, the synthesized controller enables satisfactory "velocity tracking" of the prescribed trajectory, irrespective of the perturbations acting on the system. In Fig. 3.8b is presented the positional accuracy of trajectory tracking in the longitudinal, lateral, and yaw directions for all four mentioned cases. By comparing the obtained plots it can be concluded that the synthesized control ensures a satisfactory accuracy of the nominal trajectory realization. The proposed autopilot control scheme eliminates the initial position error and guarantees satisfactory dynamic behavior of the road vehicle during motion. It can be noticed that for the selected set of control gains K_P and K_V, the dynamic vehicle behavior is not the same for the different types of perturbations. So, the "non-adapted" increasing speed on a slippery road deteriorates more significantly the quality of trajectory tracking as compared with the influence of the wind gust of moderate intensity. A slippery road and increasing forward speed during cornering have a pronounced adverse effect on the system behavior in the lateral and yaw directions. Besides, the synthesized controller ensures the motion controlability (Fig. 3.8b), whereby it should be noted that the controller is realizable only for the case when the system parameters are estimated with a high degree of accuracy.

Concerning the accuracy of the vehicle body position in the vertical direction and accuracy of the chassis pitch and roll angles, one can draw the following conclusions. By comparing the results given in Fig. 3.9 the transient regimes

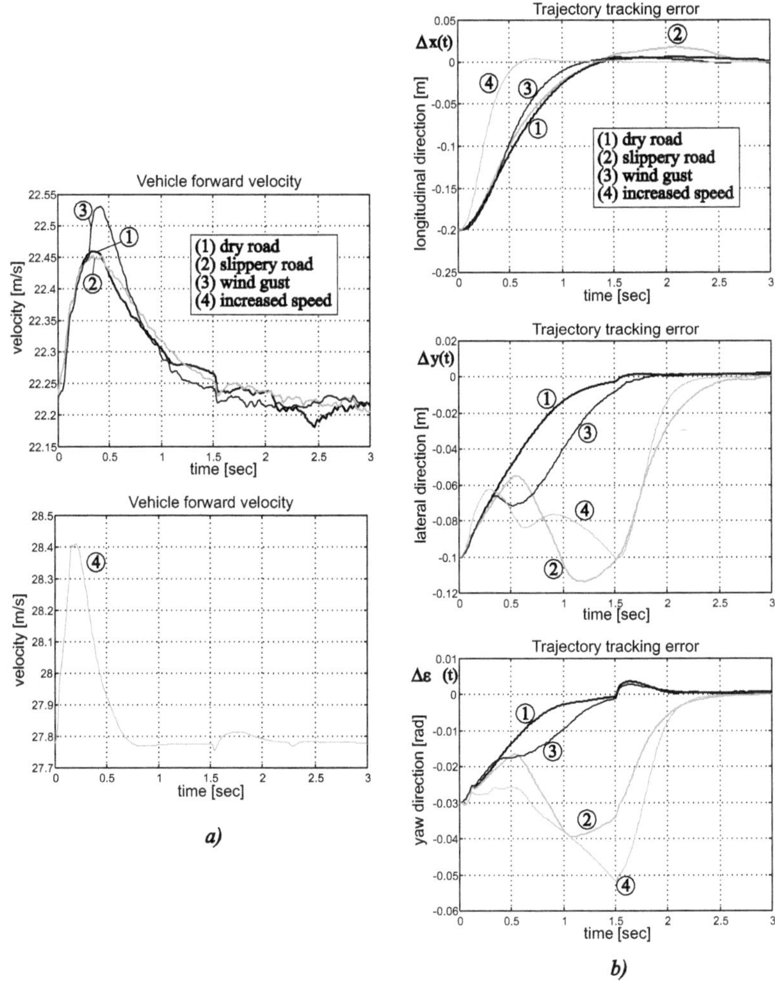

Figure 3.8: Simulation results: a) vehicle forward speed during motion under various disturbances, b) accuracy indicators for the programmed trajectory tracking in longitudinal, lateral and yaw directions

of motion of three state quantities z, ϕ and θ are essentially worse in the case without automatic control. The indicators of the errors $\Delta z(t)$, $\Delta\phi(t)$, $\Delta\theta(t)$, as well as the SS deflections for instance for the wheel no. "1" (see Fig. 2.13) are presented in Fig. 3.9. As for the ride comfort quality, the indices of vehicle dynamic behavior in the direction normal to the road surface were analyzed. Aiming at that, we introduced the so-called "average square relative error" of the observed variable expressed in %. Let us denote by ξ_i some of the following variables P_i, F_{zi}, e_i and e_{ti}. They are denoting successively the vertical payload of the i-th joint of SS, the vertical tireload of the i-th wheel, SS deflection and tire pneumatic deflection. Then, the average relative square error is defined as:

$$\sigma_{\xi_i} = \frac{\sum_{j=1}^{N}\left(\frac{\xi_i^j - \xi_{i0}}{\xi_{i0}}\right)^2}{N} \times 100 \ [\%], \tag{3.17}$$

where ξ_{io} is the nominal value of ξ_i and N is the number of samples. The values of the indices σ_{P_i}, $\sigma_{F_{zi}}$, σ_{e_i} and $\sigma_{e_{ti}}$ are systematized in Table 3.1 and given for the example of motion on an uneven road surface with and without active suspension of the wheels.

These indices represent the measures of force and acceleration amplitudes in the vertical direction, as well as the measures of deflections of the SS and tire pneumatics. By comparing the values in Table 3.1, it can be clearly noticed that the effects of bobbing due to the ride on an uneven road surface are many times lower in the case of active suspension control than in the case of a passive SS. Therefore, the ride comfort has been significantly improved. Vertical vibrations (accelerations) were lowered by the action of the active damping force f_{a_i} in the viscous cylinders of the active suspension system. Using the control scheme from Fig. 3.1, relatively small tire deflections $e_{ti} = z_t^i - z_r^i$ (see Fig. 2.29) were also ensured. It should be noted that simulation did not encompass signal prediction concerning the road surface variations.

By using the relations for calculation of the uniform distribution of control actions onto all four wheels (3.6), the proposed control scheme guarantees the minimum variations of the control variables $\delta(t)$, $\omega(t)$, and $f_a(t)$. In this way, a lower energy consumption, more uniform dynamic loads of system actuators, as well as better vehicle maneuverability are achieved. The above control variables, realized at the actuators outputs are presented in Fig. 3.10 for the case of an uneven and partially slippery road.

The presented results demonstrate that the proposed controller does significantly improve the motion stability and dynamic system performance under perturbations of various nature and intensity.

3.3.4   Essentials and Further Research Directions

In the text above we presented an integrated controller of road vehicles. The control strategy is based on the knowledge of spatial, nonlinear model of the system. It is supposed that the estimation of unknown model parameters was performed with high accuracy. Control strategy is based on the hierarchical

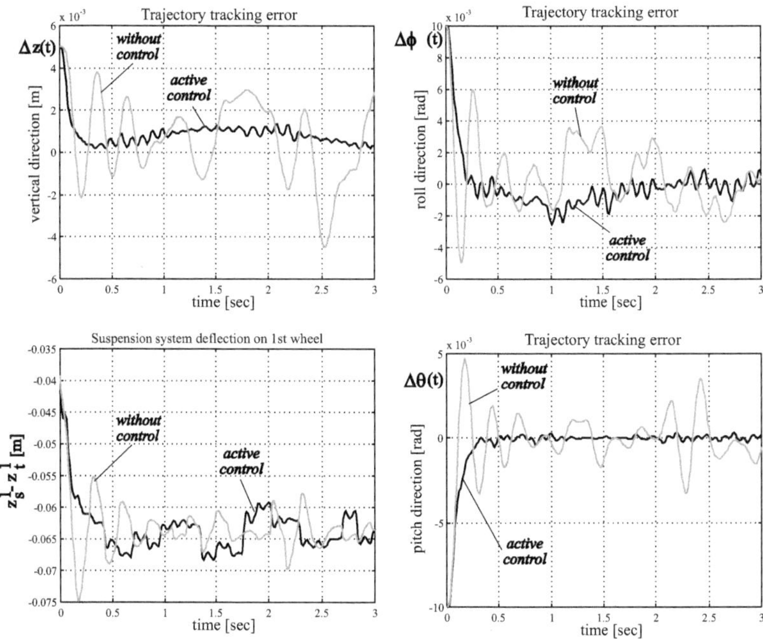

Figure 3.9: Simulation results - Accuracy indices of vertical position, roll and pitch angles; Amplitudes of the SS deflection

principle of centralized, dynamic, distributive control, so that the higher level control signals are uniformly distributed onto the actuators of particular active systems on the lower level. The obtained simulation results indicate high quality of the control performance and system stability. Here we are going to run back over the proposed strategy of dynamic control of road vehicles and consider critically the results obtained in the simulation experiments.

- **Question of the accuracy of dynamic model structure.** The nonlinear system model implemented in the proposed control scheme represents a model which describes the system's behavior in a wide range of variation of input variables and disturbance quantities. Time delays in the servoactuators, representing internal system disturbances, are not encompassed by this model. These delays have to a certain degree a negative effect on the quality of the system's behavior. Such and similar effects can be compensated for by combining conventional and unconventional control methods, and before all, by the synthesis of an additional neuro-

Table 3.1: Average relative square deviations of forces P_i, F_{zi} and deflections e_i and e_{ti} from their nominal values

			σ_{P_i} [%]		$\sigma_{F_{z_i}}$ [%]		σ_{e_i} [%]		$\sigma_{e_{ti}}$ [%]	
without control (passive sys.)	1.	2.	5.7482	5.7371	4.5536	4.5659	0.5618	0.4182	1.2428	1.3190
	tire									
	3.	4.	6.9911	7.2442	5.5632	5.8477	0.4442	0.4982	1.8607	1.9220
with active control	1.	2.	0.1450	0.0694	0.7735	0.5516	0.5396	0.3588	0.2294	0.1087
	tire									
	3.	4.	0.0201	0.0687	0.8324	1.2218	0.4641	0.6551	0.1450	0.2013

or fuzzy-compensators. They possess the capacity of compensating the effects of non-modeled system dynamics, time delays of the actuators and other structural and parametric uncertainties in the system.

- **Question of the accuracy of estimation of dynamic vehicle parameters and parameters of tire-road interaction.** This issue depends to a great extent on the available equipment for measuring the state variables of a real system or the synthesis of the corresponding state observers. Some of the values of input quantities such as road surface elevation angles (front and side) are unmeasurable or difficult to be measured from the vehicle. This information, as well as the data about the road geometry, must be sent to the controller from the outside or to be observed in some way. Also, here arises the problem of filtration of the measured signals, especially of the measured accelerations and forces, containing useless information as noise.

- **Problem of predicting signals of the disturbances acting on the system.** It is obvious that diverse external and internal disturbances can act on the system. Some of external disturbances can be determined (measured) in advance, prior to their action on the system. Such signal prediction can be carried out by measuring in some way the change of amplitude of the road surface profile.

The control algorithm based upon the knowledge of the overall deterministic vehicle model is capable of influencing simultaneously the system's behavior in the main six coordinate directions of motion. At that, there exist the possibility of improving control quality by amending the model (or its parts) on the basis of the results obtained by experiment. For example, instead of a nonlinear tire model [26, 27] used in the synthesis of our controller, it is possible to use the Pacejka's empirical model of tire pneumatic based on the experimental measurements [28, 29]. Besides, the proposed control strategy based on the application of dynamic controller represents a sound basis for a further upgrading by combining it with some of unconventional control techniques (generic algorithms, fuzzy logic, etc.).

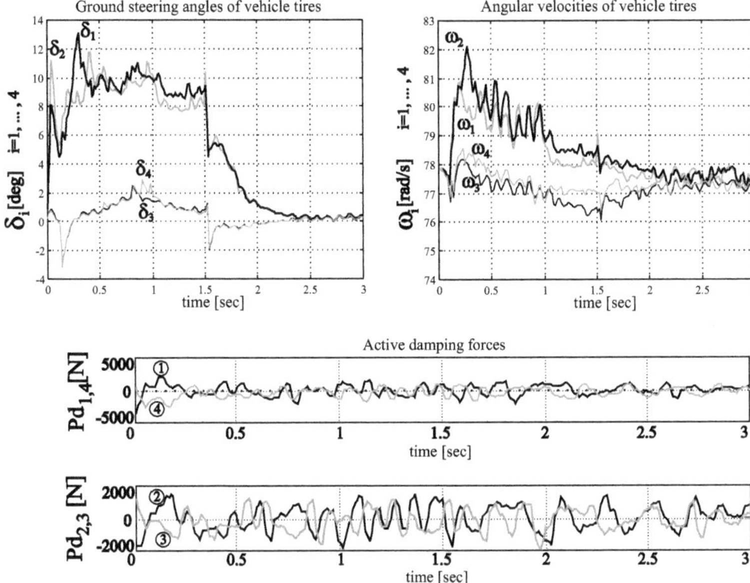

Figure 3.10: Output signals of vehicle actuators for the case of a slippery road

In the text to follow, we shall focus our attention on the way of improving the quality of the developed control algorithms (3.3) and (3.8) with the aim of adapting them to real functioning conditions. In that sense we will present the synthesis of a control that will be robust with respect to the system's parametric and structural inaccuracies, with the possibility of tuning control parameters of the controller on the basis of the analysis of practical system stability. Also, we will describe the synthesis of an additional neuro-compensator capable of greatly eliminating the negative effects of the delays in the system and effects of inaccuracies in the model structure.

3.4 Synthesis of the Neuro-compensator

In the preceding section we described in detail the centralized dynamic controller dedicated to automatic control of road vehicle performance simultaneously in the main six directions of motion. The proposed controller ensures stable motion and desired quality of vehicle behavior only in the case when the mathematical model of the object describes sufficiently accurately the dynamic behavior of the system and on the assumption that the deviations of model parameters from their real values are small. In the controller synthesis, it was assumed that the adopted spatial vehicle model, presented by the vector relation (2.28), describes relatively accurately the dynamic behavior of the system in the considered six coordination directions of motion. Of course, the applied model, because of its limited complexity, does not describe all the dynamic effects that can appear in the system. This is primarily related to the presence of elastic modes of the vehicle mechanism, presence of the Coulomb's and viscous friction in the system, dynamics of executive organs, delays in the power-steering subsystems, hysteresis in the elastic SS elements, etc. Also, the control quality is significantly influenced by the error of measurement and estimation of the state quantities appearing in the system model. All these phenomena can to a smaller or greater extent influence the overall behavior of the system. In order to enhance the robustness of the vehicle control system with respect to structural and parametric uncertainties in the system, and adapt it as much as possible to the real exploitation conditions, it is necessary to "upgrade" somehow the existing control structure of the dynamic controller. In that sense we shall propose the so-called hybrid, neuro-dynamic controller. It consists of a neuro-compensator which is added to the synthesized dynamic road vehicle controller. Under the notion of neuro-compensator we shall assume a compensator structure which is based on the artificial neural network (ANN) with four layers of perceptrons.

The task of the neuro-compensator is to compensate for the effects of model inaccuracies (both of the model structure and parameters) as well as the effect of non-modeled phenomena (e.g. actuators' delays). Neural networks (NNs) appeared to be a very useful means as they are good identifiers of nonlinear functions [30]. NNs have already been used in control structures of similar dynamic systems. In this section we are going to describe in brief the structure and application mode of the additional neuro-compensator which functions in co-operation with the synthesized dynamic controller. The synthesized hybrid, neuro-dynamic vehicle controller is based on the use of the object's mathematical model and application of a neuro-compensator obtained by training the chosen NN. The newly-proposed hybrid controller represents a combination of the control algorithm based on the knowledge of vehicle dynamic model and the knowledge accumulated in the ANN structure itself.

3.4.1 Learning of the Vehicle Model Using ANN

In the recent ten years NNs have been successfully used in the control of complex dynamic systems, taking advantage of their capability of learning and generalizing knowledge. Nevertheless, there are still some difficulties in designing control systems based on the application of NNs because these structures, looked from outside, represent "black boxes". In the majority of cases their application has not been based on analytical models derived on the basis of a real system. Here, we are going to present a new vehicle control system based on the integration of the control law known in the theory as Inverse Torque Method or Computing Torque Method and the application of NNs. The inverse dynamics method has been theoretically derived from the expressions of the dynamic model of the considered system, given by the relation (2.28). This control law was previously described by the relation (3.3). The model (2.28) represents a relatively good approximation of the real system behavior in a wide range of change of the global vehicle state variables and input variables. On the other hand, an NN can additionally identify the dynamic effects that were not encompassed by the mathematical model (2.28) and by the control law (3.3). The unmodeled system dynamics is identified by training of the selected NN structure using some of the learning methods. For this purpose the most frequently employed back propagation method was used, which has been applyed in practice to train the multilayer networks in handling similar control tasks [30, 31]. In this section, we propose one convenient way of identification of the vehicle model. Then, we shall synthesize the appropriate hybrid neuro-dynamic controller. Also, we will present the corresponding simulation results with the aim of pointing to the advantages of implementing the developed control system involving the supplementary neuro-compensator over the application of a dynamic controller alone.

The structure of the hybrid neuro-dynamic controller, whose control schemes are given in Figs. 3.1 and 3.2, consists of two functional blocks. The task of the first (model-based) block is to compensate for the main dynamic effects, while the second (knowledge-based) block is based on the application of the NN for compensating the system uncertainties [30, 31]. The automatic control system of the autonomous road vehicle is designed in this monograph by integrating the dynamic control algorithm (3.3) or (3.8) and the selected NN structure, using the model (2.28). The application of ANN is performed in three stages: (i) rough off-line training, (ii) fine on-line training, and (iii) implementation of the finally tuned net in the synthesized hybrid controller.

Taking into account the experience of other authors [32, 33] as well as [34, 35], the structure of the learning process for the off-line estimation of the unmodeled system dynamics is shown in Fig. 3.11. In the beginning, the appropriate input and output signals needed for the network training are selected. For that purpose, manual commands are set to the vehicle to produce variation of the input command signal such as the tire angular velocities ω_i and variation of the ground steering angles δ_i, $i = 1, \ldots, 4$. Apart from these, various external disturbances can act upon the vehicle during its motion. These are variations of the road surface profile, changes of intensity and direction of the side wind gust, as well as variations of Coulomb's friction coefficients between the tire pneumatics and road surface, etc. Using simulation experiments[4], the global state variables $(q(t), \dot{q}(t), \ddot{q}(t))$ are determined and saved. These state variables serve as input signals to the rough off-line training of the selected ANN. These variables are used to compute the vehicle mathematical model too, as shown in Fig. 3.11. On the basis of the values of generalized forces $\hat{\tau}(t)$ calculated from the model and their real values $\tau(t)$ obtained by measuring at each time sample on the real system during the ride, the deviation vector $\Delta\tau(t) = \tau(t) - \hat{\tau}(t)$ can be calculated. At the same time, the so-called compensation vector of

---

[4]A nonlinear vehicle model is used to obtain the necessary system input/output variables to be used for training of the selected ANN structure.

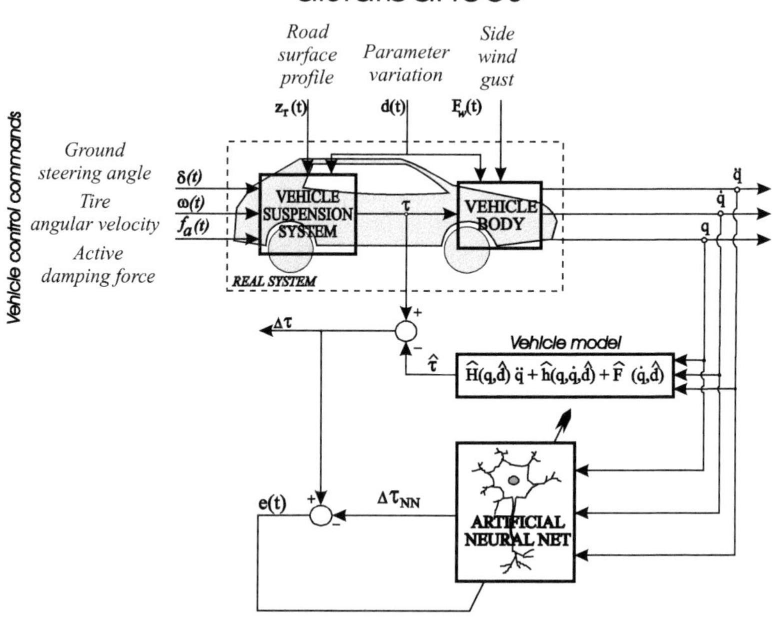

Figure 3.11: Structure of the training process of the selected ANN

forces/moments $\Delta\tau_{NN}$ is generated at the ANN output layer. Values of the (6×1) vector $\Delta\tau_{NN}$ correspond to the values of the estimated magnitudes of the forces and torques that have not been taken into account in the vehicle dynamics modeling. The obtained vectors $\Delta\tau(t)$ and $\Delta\tau_{NN}(t)$ are compared at the discriminator $(e(t) = \Delta\tau(t) - \Delta\tau_{NN}(t))$. The existing error e(t) is used for further tuning of the ANN weighting matrices (Fig. 3.11). The network is trained in a larger number of learning epochs. One epoch represents the period needed by the selected network to train in one passing all of the pairs of ordered vectors $U(t) - Y(t)$ of input-output samples within the given time interval $t \in (0, T]$, where T represents the total time of motion duration. When the amplitude of the learning error $e(t)$ is smaller than the predetermined desired value $(e(t) \leq E)$, the rough off-line training process is terminated. The network should be trained on a limited number of characteristic representative samples (packed into input vectors). The objective is to allow the selected network exhibit the properties of knowledge generalization and thus encompass a wider set of possible vehicle trajectories defined by the vectors $q(t)$, $\dot{q}(t)$ and $\ddot{q}(t)$.

Otherwise, the considered network will be a good approximator of the non-modeled system dynamics only on a relatively small set of the given trajectories. In this respect, it is necessary to have the appropriate experience and skill to carefully select a representative set of input quantities intended for training the selected NNs. The next step in the synthesis of the neuro-compensator is the so-called "fine net tuning". The weighting factors in the corresponding matrices of the suitably selected net architecture, defined by the off-line learning process, serve as initial values for further "fine tuning" of the NN. This procedure should be carried out in the real time on the real system in the closed-loop mode (Fig. 3.1). When the desired quality of learning process is attained, "fine tuning" is ended.

In Fig. 3.1 is presented the control scheme of the proposed system of automatic road vehicle control with the supplementary neuro-compensator in its structure. Structure of the hybrid, neuro-dynamic controller presented in Fig. 3.1 is split into two parts. One part of this structure is to compensate for the basic dynamic effects which appear in the system motion, while the other, based on the application of ANN, compensates for the effects of the uncertainties arising in the system. At that, the NN should be trained to such an extent that the system with a high degree of accuracy approximates the difference in the dynamic behavior of the real system and its model. The quality of compensating for the consequences arising from the differences between the behavior of the real system and its model will be the higher the better is the network identification of the mentioned structural inaccuracies and parameter uncertainties of the model.

The roughly estimated weighting factors in the network matrices serve as input values for further fine tuning of the network within the designed control system. The procedure is performed on an experimental system in real time. The next step in adjusting the neuro-compensator is a fine tuning of the weighting values of the network matrices. This is carried out on a closed-loop real system (see Fig. 3.1), whereby the dynamic controller performs its part of the control task. At that, in each particular learning epoch, the current values of the weighting factors in the network are continuously changed. When the desired quality of accuracy (training) of the net is attained, further on-line training is stopped. The values of the weighting matrices that were calculated last, represent the constant values of the newly-tuned neuro-compensator. Thus the training process is completed and the network is "ready" to resume the allocated task of additional compensation of the unmodeled system dynamics. As such, it functions within the hybrid neuro dynamic vehicle controller (Fig. 3.1).

3.4.2 Topology of the Applied Neural Network

Analytical expressions describing the structure of the network training algorithm can be found in the abundant literature from this area, e.g. in [35], [36] and [37], so that we are not going to deal with the details. The ANN architecture used in the proposed hybrid controller (Fig. 3.1) is presented in Fig. 3.12. Topology of the proposed ANN intended for associative mode of learning of the road vehicle

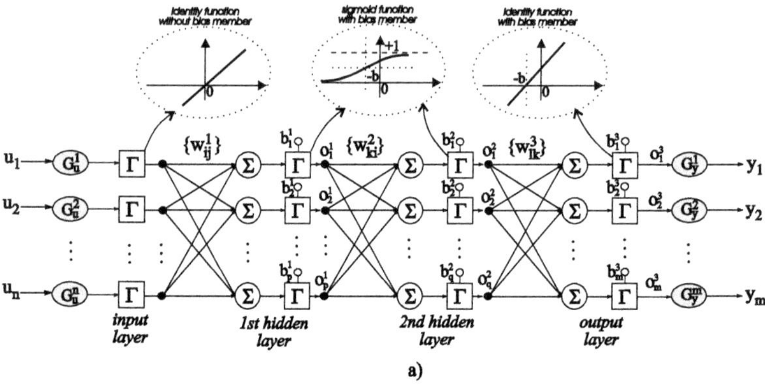

a)

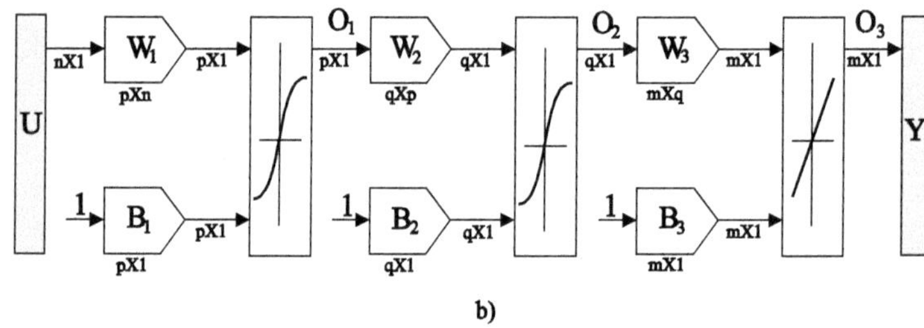

b)

Figure 3.12: a) Topology of the four-layer neural network for compensating inaccuracies and uncertainties of the road vehicle model, b) Block-diagram of layered structure of the selected network

model and its control is defined by the four layers: one input, one output and two hidden layers (Fig. 3.12). The hidden layers are introduced for the purpose of realization of the nonlinear function of mapping from the network input to its output. The layered structure of the network with the nonlinear transfer functions Γ (Fig. 3.12a) makes it possible the network learns both the linear and nonlinear relations between the input U and output Y vectors (Fig. 3.12b). As activation function in the two hidden layers we selected a sigmoidal function of the form $f(x) = 1/(1 + exp(-x))$, where x denotes input signal to the transfer function Γ. This signal represents the sum of the weighting input signals of one perceptron (Fig. 3.12b). Only in the input and output layers use was made of the identity function (linear function) as an activation function. The linear output layer allows the network to generate at its output the values outside the range $[-1, +1]$. For this reason, at the network output (and formally at its input too) there are the so-called matching gains G_u^i and G_y^i. Their role is to match dimensionally the values of the output vector with the values of the input vector. This is necessary in order to attain a better and faster network convergence, i.e. to accelerate the training process. In addition to the mentioned details related to the network structure, the output and hidden layers contain bias members. The corresponding "bias-vectors" of particular layers, denoted by B_1, B_2 and B_3, are presented in Figs 3.12a and 3.12b. The role of these bias members is to give the network more DOFs in learning the given function. The use of the bias members results in better training effects, i.e. it enhances the prosepects that the information by back propagation in the network attains finally the acceptable solution, diminishing thus the number of epochs needed for training. Bias members are given as units (Fig. 3.12b) and, are being constantly changed in the course of training the corresponding weighting values included into the bias-vectors, such as weighting values of the matrices W_1, W_2 and W_3 (Fig. 3.12b).

The ANN presented in Figs. 3.12a and 3.12b has an input layer of $n = 15$ perceptrons and the output layer $m = 6$ perceptrons. These dimensions are dictated by the nature of the technical system having 6 DOFs in the main coordinate directions of motion (Fig. 2.13). As the relative position of the vehicle body MC considered with respect to the fixed coordinate frame attached to a point on the road surface does not influence the vehicle dynamic behavior during motion, the coordinates $q_1 = x$, $q_2 = y$ and $q_4 = z$ are omitted and are not taken as elements of the vector U. Because of that the vector U is of dimension (15×1) and it has now the following elements:

$$U = [q_3 \ q_5 \ q_6 \ \dot{q}_1 \ \dot{q}_2 \ \dot{q}_3 \ \dot{q}_4 \ \dot{q}_5 \ \dot{q}_6 \ \ddot{q}_1 \ \ddot{q}_2 \ \ddot{q}_3 \ \ddot{q}_4 \ \ddot{q}_5 \ \ddot{q}_6]^T \qquad (3.18)$$

where q_i, $\dot{q}_i$ and $\ddot{q}_i$ are the elements of the corresponding $(n \times 1)$ vectors of the position q, velocity $\dot{q}$ and acceleration $\ddot{q}$. Elements of the network output vector Y are the corresponding forces and moments acting at the vehicle body MC, along and about the axes in the particular directions of the object motion:

$$Y = [\Delta \tau_{NN}^1 \ \Delta \tau_{NN}^2 \ \Delta \tau_{NN}^3 \ \Delta \tau_{NN}^4 \ \Delta \tau_{NN}^5 \ \Delta \tau_{NN}^6]^T \qquad (3.19)$$

where the terms $\Delta \tau_{NN}^i$ $(i = 1, \ldots, 6)$ have dimension of forces/moments acting at the vehicle body MC. One of the main parameters of the synthesized ANN represents the number of perceptrons in each layer. In order to determine the effect of the number of perceptrons in the network hidden layers on the performance of the learning algorithm, experimental training was carried out with different numbers of perceptrons in the particular hidden layers. On the basis of simulation experiments the following optimal network topology was determined: $n = 15$ perceptrons in the input layer, $p = 81$ perceptrons in the first hidden layer, $q = 67$ perceptrons in the second hidden layer, and $m = 6$ perceptrons in the output layer. Looking at Fig. 3.12b it is possible to give final dimensions of particular vectors and matrices of the selected NN. The weighting matrices in the net are of the following dimensions: $W_1(81 \times 15)$, $W_2(67 \times 81)$ and

$W_3(6 \times 67)$. Dimensions of output vectors in the particular network layers are: $O_1(81 \times 1)$, $O_2(67 \times 1)$ and $O_3(6 \times 1)$. Dimensions of the bias vectors B_1, B_2 and B_3 correspond to the respective dimensions of the corresponding output vectors of the layers O_1, O_2 and O_3. When all the dimensions of the corresponding vectors and matrices in the NN structure are known, it is possible to define the concrete values of the "matching gains" G_u and G_y which are experimentally estimated on the basis of simulation studies. The adopted values of these gains are:

$$
\begin{aligned}
G_u^i &= 1, \quad \text{for } i = 1, \ldots, 15 \\
G_y^j &= 0.01 \quad \text{for } j = 1, \ldots, 6
\end{aligned}
\tag{3.20}
$$

The way of training of the network defined above, used in the estimation of certain parameters of road vehicle control, will be demonstrated on a simulation example described in the subsequent section.

3.4.3 Simulation Experiment with the Neuro-compensator

In this section we will show how the application of the additional neuro-compensator in co-operation with the vehicle dynamic controller can improve the overall dynamic behavior of the system compared with that observed when using only the dynamic controller. Also, we shall demonstrate the possibility of improving control performance of the proposed system using the property of associative learning with NNs. Hereby, our objective is not to find out optimal solutions to the structure of multilayered networks suitable for the application in the hybrid controller. Our intention is only to show the way how a multilayered NN can be integrated into the structure of the hybrid neuro-dynamic controller of road vehicles and what are the effects it produces on the quality of object control. In order to demonstrate the advantages of using NNs in the considered control tasks we chose a characteristic example of control of a road vehicle moving along the given nominal trajectory. The shape of the desired trajectory is shown in Fig. 3.13. The spatial nonlinear model given by the relations (2.28) - (2.54), whose parameter values are given in Appendix of Chapter 2, was used in the simulation experiment. At that, the purely positional control law (3.3) was chosen. The corresponding gains were synthesized to ensure the system's stability in all directions of motion over the considered time interval. We adopted the values of position and velocity gains K_P and K_V used also in the previous simulation example, whose values are given in the subsection 3.3.3.1.

To determine the weighting values of the matrices in the selected NN shown in Fig. 3.12 use was made of standard back propagation algorithm. For the purpose of network training we prepared pairs of input-output vectors obtained by simulation of the vehicle model (2.28)-(2.54). At that, different control commands as a function of time were given to the vehicle (see Fig. 3.14), i.e. an open-loop system operation was simulated. The assigned commands comprised continuous changes of the ground steering angles $\delta_i(t)$ $(i = 1, \ldots, 4)$ of the tires

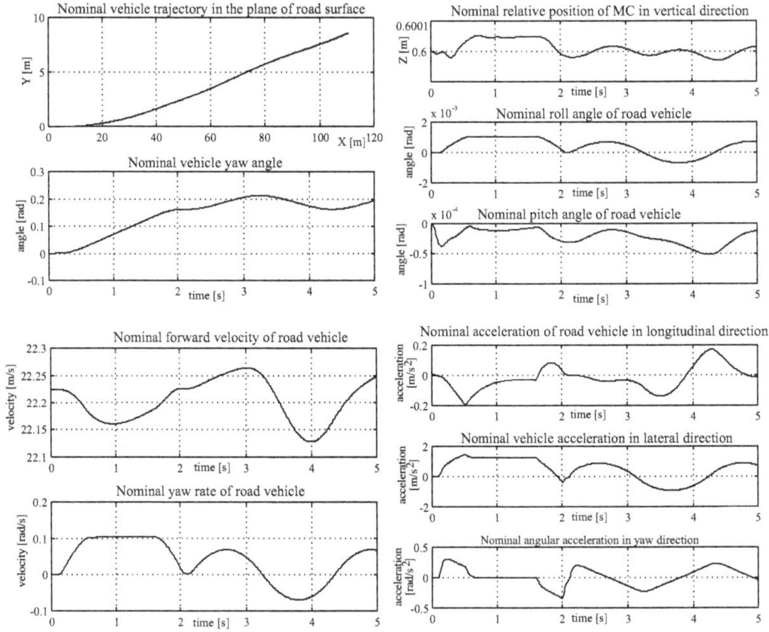

Figure 3.13: Nominal trajectory used in simulation of the vehicle model with the hybrid neuro-dynamic controller

and changes of the tire angular velocities ω_i ($i = 1, \ldots, 4$). Also, the effect of certain external random disturbances acting on the system was simulated. In order to be able to perturb the "vertical" dynamics too, in addition to the perturbation of the longitudinal and lateral dynamics, it was assumed the occurrence of variations of road surface profile on all four wheels. Equivalent effects on the system vehicle model could be obtained by assigning random changes of the amplitudes of active suspension forces f_{a_i} ($i = 1, \ldots, 4$) in the viscous cylinders of the SS (Fig. 2.29). The maximal variation of road unevenness was chosen to be in the range of $z_r^{max} = \pm 2.5$ [mm]. The above commands and disturbances were given in a total time interval of 100 [s]. The interval of recording current values of position $q(t)$, velocity $\dot{q}(t)$ and acceleration $\ddot{q}(t)$ of the vehicle (in the main six coordinate directions) was chosen to be $\Delta t = 0.04$ [s]. In Fig. 3.14 are presented the input quantities of the vehicle model used, selected for generating the values of the NN input vector $U(t)$. The given ground steering angles of front tires shown in Fig. 3.14, including abrupt changes of input quantities of different amplitudes were 3.5 and 5.5 [deg]. The two sinusoidal inputs of the

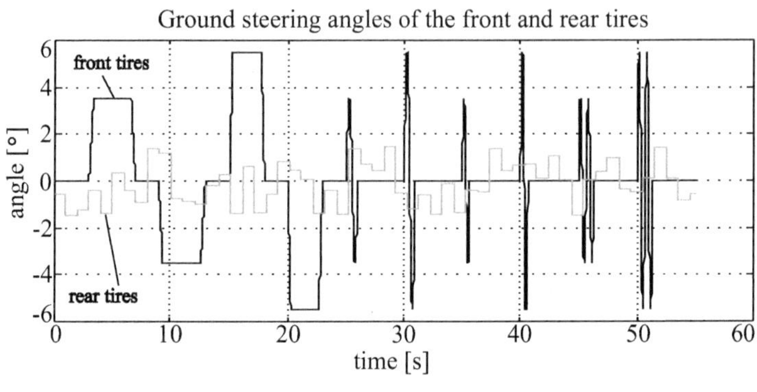

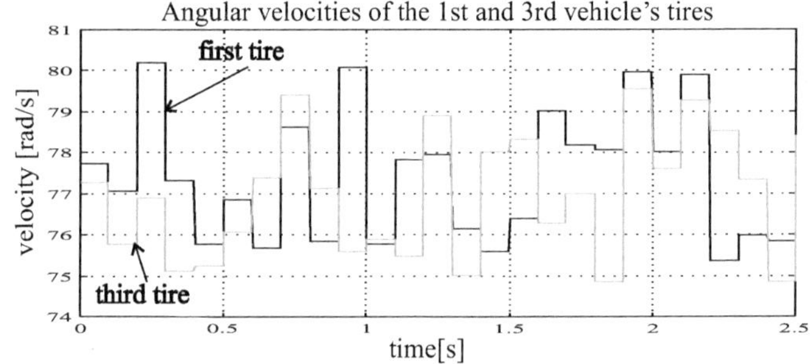

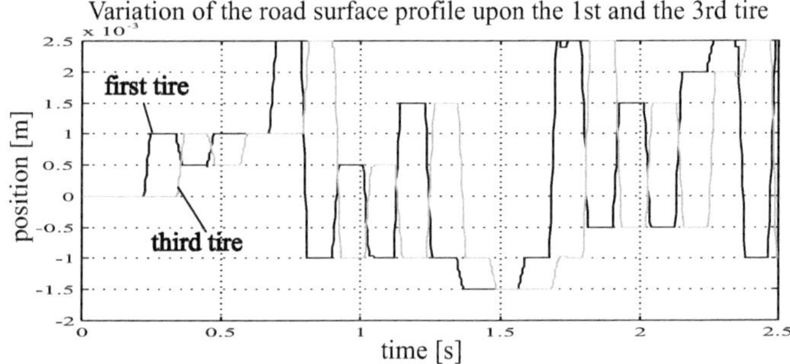

Figure 3.14: Excitation commands and assumed disturbance signals to generate pairs of input-output vectors in the NN used in the identification of the vehicle dynamic model

same amplitudes of 3.5 and 5.5 [*deg*] and frequency of 1 and 1.5 [*Hz*] were also given. On the other hand, it was supposed that rear tires angles are defined as random pulses whose amplitude is not exceeding 1.5 [*deg*]. The network was trained under conditions of vehicle motion with a linear forward speed of $V = 80$ [*km/h*] on a dry road with variations of the amplitudes of angular velocity of wheel shown in Fig. 3.14. The total number of items used in the network training was 2500. They were the result of sampling over a time interval of 100 [s] with a sampling time of 0.04 [s].

As mentioned above, the input vector $U(t)$ represented by the expression (3.18) and intended to train the NN contains the terms representing the corresponding relative positions, velocities, and accelerations of the vehicle body in the particular directions of motion. The stored values of the network input vector in particular time instants t are used in the off-line training regime. The hodographs of the particular state quantities in the period of first 2.5 [s] of vehicle motion are given in Fig. 3.15. These quantities form the network input vector $U(t)$. For the sake of eligibility, the graphs do not show the considered quantities over the whole time interval of motion duration $t = (0, 100]$ [s].

On considering Fig. 3.1 it comes clear that the aim of the network training was to achieve that, after completing the procedure, the network is capable of reproducing the values of forces and moments $\Delta \tau_{NN}(t)$ corresponding to the forces and moments which would be sufficient to compensate for the effect of uncertainties and inaccuracies of the road vehicle model. Simply speaking, the network represents an approximate nonlinear model of the vehicle behavior that is not encompassed by the analytic model (2.28). The idea is that the NN presented in Fig. 3.12, on the basis of the acquired knowledge of unknown vehicle dynamics, compensates for the non-modeled effects and effects of parameter inaccuracies on the system behavior. Related to this, in the simulation example we assumed the existence of some structural and parametric inaccuracies of the model. In this way, the real conditions of system functioning were simulated. As we have already pointed out, in practice one can never generate an ideally accurate model of a system, especially of a complex nonlinear system such as road vehicle. As for the structural model inaccuracies used to train the network to identify them, it was supposed that in the vector F (represented by the expression (2.32)) of the model (2.28) the term representing the model of forces and moments of rolling resistance was unknown. Thus, we simulated the inaccuracy of the structure of the mathematical model used for the synthesis of dynamic control. Apart from structural inaccuracies, some parameter uncertainties of the model were considered. The task of training the selected network was to identify these inaccuracies (parametric and structural), to be able later to compensate for their effects in the frame of the proposed control law. In this way is enhanced the robustness of the control system with respect to the presence of model inaccuracies. In the stage of preparing data intended for network training, simulation of parameter inaccuracies was performed. Thus, the elements of the vector of system parameters d (expression (2.28)) were made to vary in the range of ± 5 %. It was supposed that the errors of estimation of parameters of the system model could not exceed the predicted limit.

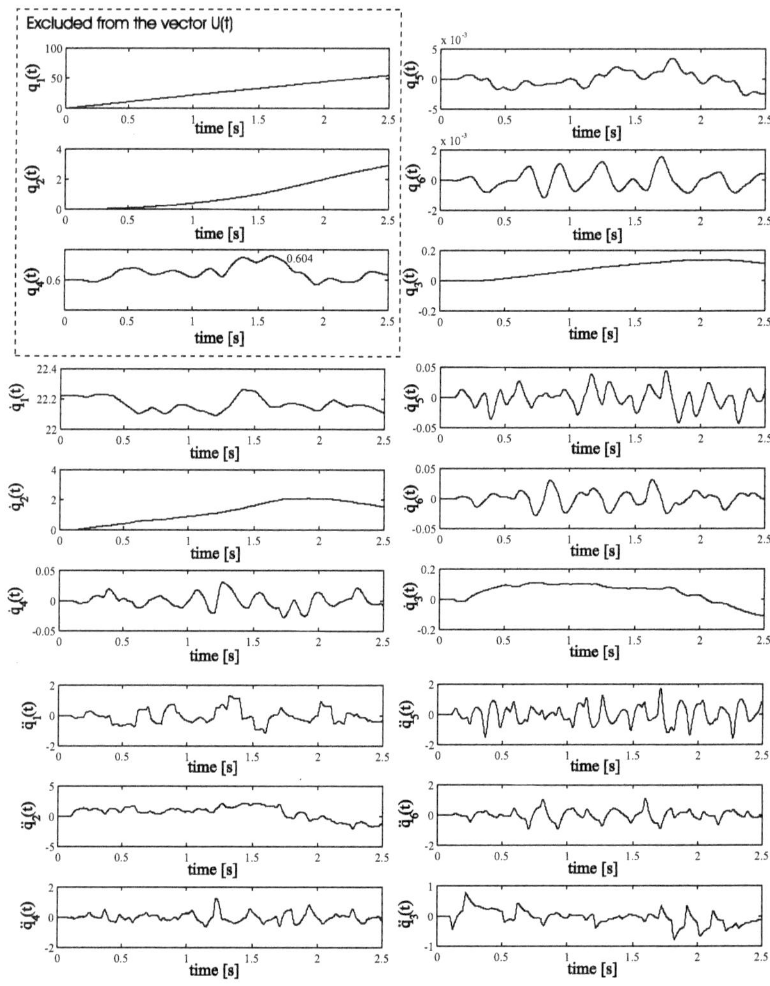

Figure 3.15: Hodographs of the quantities $q(t)$, $\dot{q}(t)$ and $\ddot{q}(t)$ over a time interval of $t = (0, 2.5]$ $[s]$ as the quantities consisting the input vector (3.18) intended to train the chosen NN

The network training was carried out using the programs from MATLAB "Neural Network Toolbox" [38]. Prior to training we defined the network training parameters: target error $E = 180$, maximal number of learning epochs $e_{max} = 10^5$, and the initial value of learning rate coefficient $l_r = 0.01$. The value of the learning rate coefficient was determined by experiment. At higher values of the coefficient l_r it was not possible to attain a satisfactory degree of error convergence $e(t)$ (Fig. 3.16), or the degree of convergence was poor. The way how particular indicators change during the network training process is essential for estimation of the quality of training. This is illustrated in Fig. 3.16. The network training is completed when the sum of the squares of errors of the output vector $Y(t)$ drops below the prescribed value $e(t) \leq E$. Then, the current values of weighting matrices W_i and elements of the network bias vectors B_i $(i = 1, 2, 3)$ are saved. The weighting matrices and bias vectors thus determined are the result of the rough off-line network training. After that comes the next step, the so-called fine tuning. It is carried out in a closed-loop system, using a larger number of newly-assigned nominal trajectories. The criterion of the network training efficiency is a sum of the squares of errors of trajectory tracking. Values of the matrices W_i and vectors B_i $(i = 1, 2, 3)$ represent the knowledge accumulated in the network structure. This knowledge is related to the characteristics of model uncertainties present in the system.

The accuracy indices of tracking the nominal trajectory given in Fig. 3.12 are illustrated in Figs 3.17 and 3.18.

In the simulation example a side wind gust was introduced. In the time instant $t = 0.5$ [s], a side wind gust starts to act as an external disturbance of the system. Intensity of the wind gust impact is presented by diagrams of time dependence of the force/moment time history acting on the vehicle body during its motion. Components of the vector $F(t)$ are given in Fig. 3.19. They are presented in three coordinate directions as viewed with respect to the direction of the vehicle motion. Simulation results in Figs 3.17 and 3.18 are presented for two cases of control (with and without the additional neuro-compensator). It can be noticed that in the case of the parameter estimation error of ± 1.5 % with respect to the real parameter values, as well as for the supposed inaccuracies of the model structure (due to neglecting the rolling friction coefficient in the expression (2.32)), better results are obtained for the accuarcy of nominal trajectory tracking in the case of application of the neuro-compensator than without it. The obtained differences in control quality in the two cases are visually rather small, which is a consequence of the assumption that the uncertainties and model inaccuracies worc also small. Irrespective of this, even relatively small improvements represent a sufficient indicator of the expected advantages offered by the implemented NN integrated in the control scheme of the vehicle controller.

The control signals $\Delta \tau_{NN}(t)$ obtained at the output of the dynamic neuro-compensator during motion are presented in Fig. 3.20. These signals have dimensions of forces and moments which should additionally act at the vehicle body MC. Their role in the control system is to compensate for the identified structural inaccuracies of the system model and parameter uncertainties in using

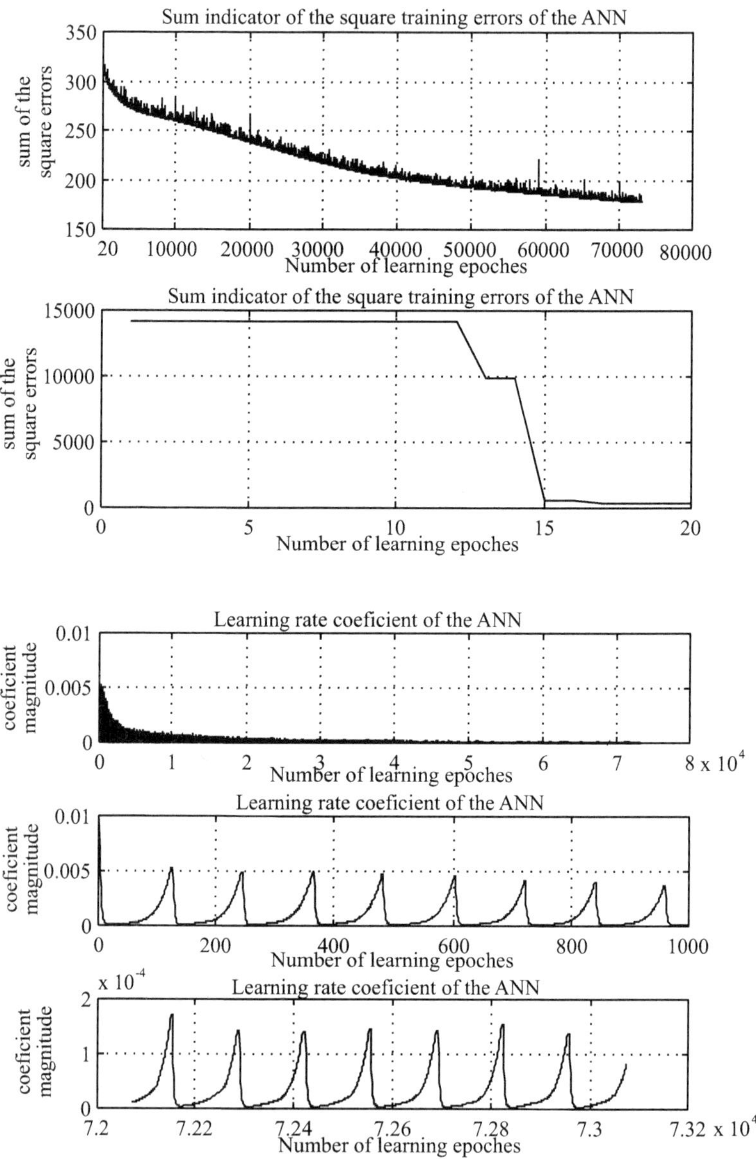

Figure 3.16: a) Degree of error convergence of the network training, b) Values of coefficients of the rate of network training as a function of the number of training epochs

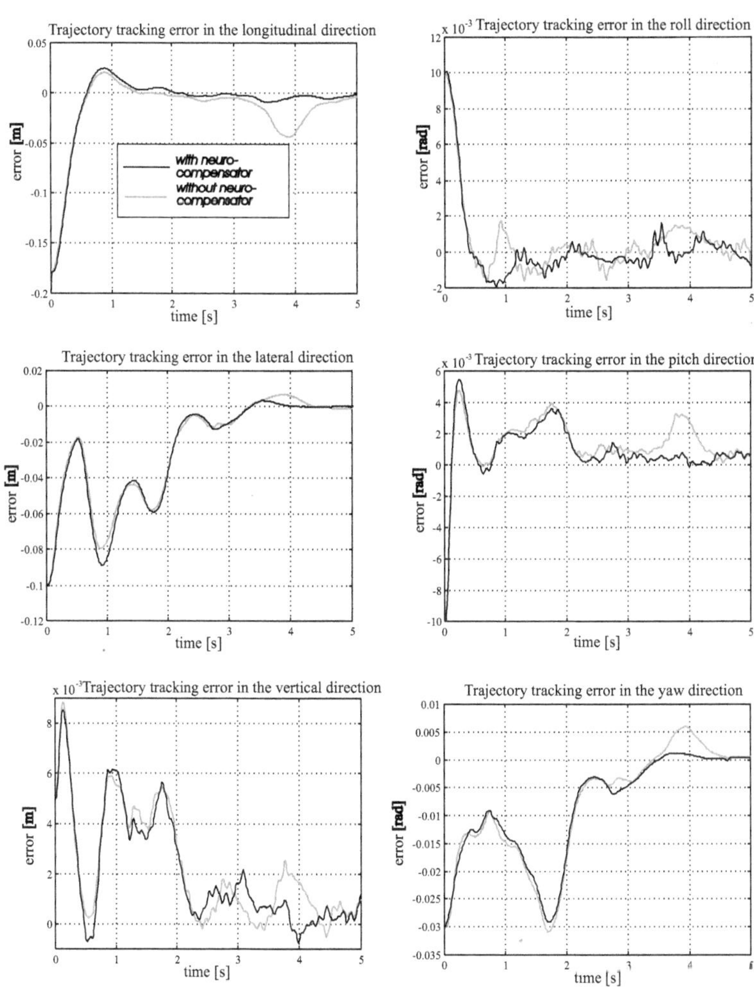

Figure 3.17: Indicators of accuarcy of tracking the nominal trajectory with respect to position in main six coordinate directions of the vehicle body motion

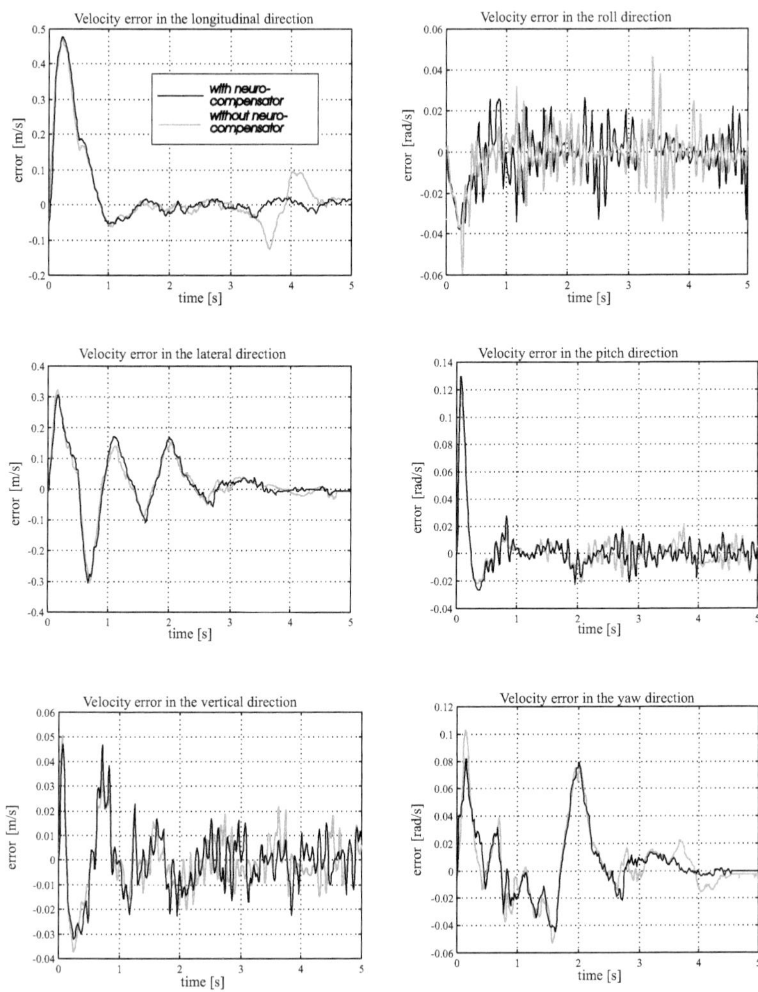

Figure 3.18: Indicators of accuracy of tracking the nominal trajectory with respect to velocity in main six coordinate directions of the vehicle body motion

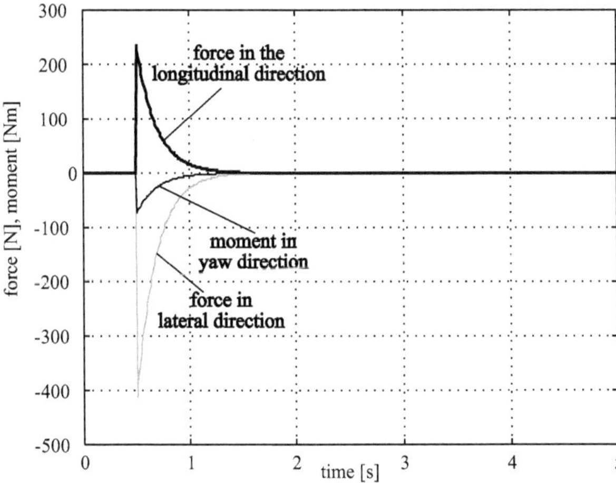

Figure 3.19: Components of the vectors of forces/moments acting on the vehicle body as the side wind gust in three directions: longitudinal, lateral, and yaw angle

the approximate model for the synthesis of the controller. The presented signals have a noise-like appearance. Hence, it would be desirable to carry out filtration of these signals prior to their integration with the signals originated from the other part of the dynamic controller (see Fig. 3.1).

3.5   Implementation Aspects of the Proposed Controller

The proposed controller was synthesized in the three stages:

(i) First, non-adaptive integrated dynamic controller of road vehicle based on the application of the pure position or combined position-force control algorithm at the tactical level was designed. The application of the dynamic controller demands real-time estimation of model parameters with a relatively high accuracy;

(ii) In the second stage, control gains were calculated on the basis of the practical stability test. In this way, a greater robustness of the control system with respect to the system uncertainties was attained;

(iii) Finally, to the existing controller architecture, a supplementary neuro-compensator, based on the application of the multi-layered NN structure, was added. It gave an additional quality to the automatic control system a higher

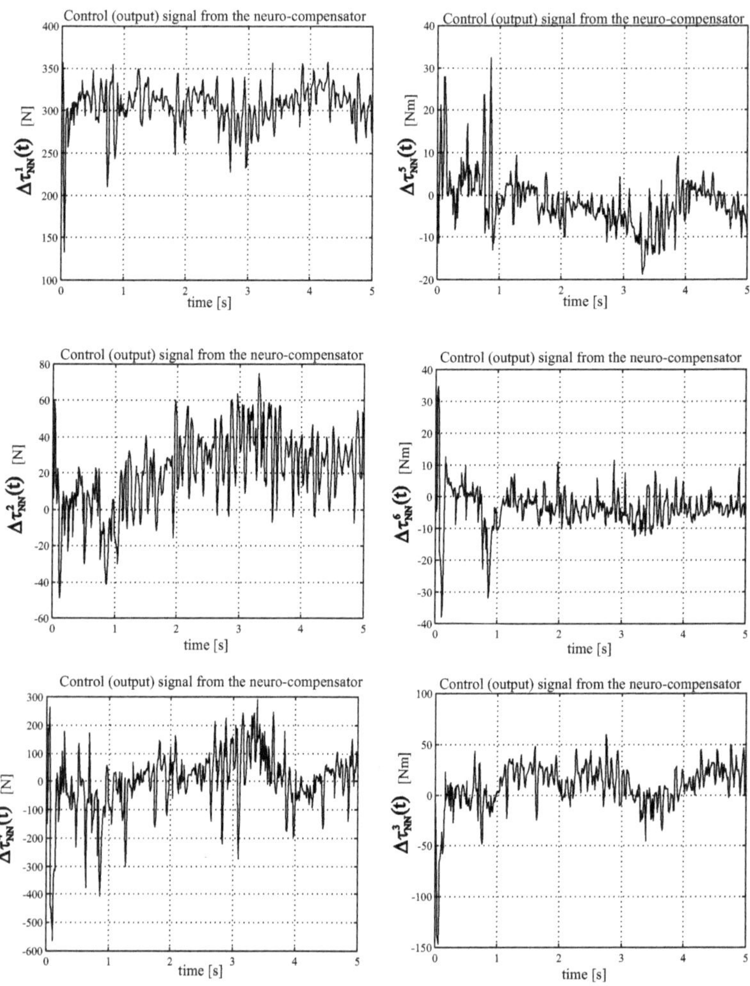

Figure 3.20: Control signals at the NN output applied to control vehicle motion along the given trajectory presented in Fig. 3.13

robustness against the system modeling inaccuracies.

On the basis of the analysis of the obtained simulation results it can be concluded that the proposed hybrid neuro-dynamic controller represents an advanced control system which ensures the stability of the system as well as desired ride quality in all six directions of motion simultaneously. Further improvements should be sought in the application of some other NN topology or in choosing a more efficient method which would enable increase in the rate of associative learning of the model used in control. In that way a higher efficiency of the network in the control process should be attained. Also, one of possible further improvements of the proposed controller would be in the decentralization of the control system, which should additionally simplify design of the controller.

In the previous sections, theoretical backgrounds of the synthesized neuro-dynamic vehicle controller was described, intended for automated vehicle control. In this section, more attention will be paid to the implementation aspects of the designed vehicle autopilot controller.

The proposed hybrid controller demands the existence of the appropriate sensor equipment and the corresponding active control systems. Concerning the necessary sensors which are indispensable for the functioning of the proposed control system, they depend on the choice of the applyed type of dynamic control law: purely positional or combined position-force control algorithm. What will be the law to be implemented it depends on the vehicle dedication (passenger car, transport vehicle, off-road car, etc.) and the behavior to be achieved in its exploitation. In the case when is necessary to keep a certain constant attitude deflection of the vehicle body during motion w.r.t. the road surface, it is more suitable to implement the pure position control law (3.3). On the other hand, if it is demanded that the vehicle minimizes variations of acceleration in certain directions (especially in the vertical plane), it is suitable to implement in the controller the combined position and force/moment control (3.8). Related to this, the demands for an appropriate sensor equipment will differ in the two cases.

The proposed controller demands design and implementation of the corresponding active system in the vehicle such as four-wheel active steering system, four-wheel active driving/braking system, and active SS. Functional schemes of these three active systems are presented in Figs. 2.29-2.31. In contrast to conventional vehicle driving and power transmission based exclusively on IC engines we propose here the use of the so-called hybrid power drives [19] on all four wheels. Such a solution is presented in Fig. 2.31 by its functional scheme. Thanks to the advantages this solution exhibits over conventional driving system it is expected that it will replace in full conventional drives in the future. The use of the hybrid drivetrain yields active traction control, i.e. independent braking at each wheel. In this way it is possible to actively control the vehicle motion both in the longitudinal and lateral directions at the same time. As for the active steering system and active suspension system, they have already been available with the new-generation road vehicles. So, the newly-proposed approach to control road vehicles assumes also new design solutions for the system actuators. In our opinion, the newly-proposed controller fits well into the development perspectives of the automotive industry. Such concept assumes a higher degree of vehicle robotization applied in the traffic, notably on automated highways, which will represent the major transit lines of the urban environment in the future.

Realization of the designed hybrid, neuro-dynamic controller, one possible version of which is given by its control scheme in Fig. 3.1, is based on microprocessor control by means of the chosen actuators (like those in Figs. 2.29-2.31). In view of the complexity of the control scheme an appropriate controller hardware architecture has to be foreseen, which would correspond to the needs of the control task. In dependence of the chosen types of vehicle actuators, which is a question of design, it is necessary to adapt the hardware structure of the controller too. In that sense a multi-processor control system has to be designed, which would be able to perform several tasks in real time. These are before all: processing of the oncoming signals from the sensors and communication equipment, real-time estimation of the system dynamic parameters and the parameters of tire pneumatics-road surface interaction, choice of the controller gains according to the variable vehicle ride conditions by selecting the corresponding gain values from the controller memory, calculation of the control signals at the tactical control level, determination of the reference signals for the lower control level, synchronization of the operations within the controller, calculation of the feedforward signals as outputs from the neuro-compensator, etc. So, due to the multitasking operation mode, the controller should be designed as a multiprocessor system, so that it simultaneously performs the following basic functions: (i) estimation of the model parameters and parameters of the vehicle-road surface interaction in real time, (ii) calculation of the global control signals at the tactical control level and, (iii) calculation of the local control actions at the level of system's actuators. It would be convenient to have a controller with four processors in its structure: master (main) processor (MP) and three slave (auxiliary) processors (P1, P2, P3). The architecture of such a controller structure is shown in Fig. 3.21.

The main processor has to coordinate the operation of the multiprocessor system and synchronize task execution within the controller (it would regulate time sequence of performing individual processes). The slave processors should operate in parallel regime, performing the given functions simultaneously. The P1 processor should perform real-time estimation of the unknown model parameters based on the available information from the sensory system and communication equipment. It sends the results of these procedures to the main processor. On the basis of the determined system parameters and parameters of the road surface characteristics, the main processor should take from the memory the corresponding set of control signals. In that way, adaptability of the control system actions to variable driving conditions would be achieved. The MP would proceed the taken gains to the P2 processor. In P2, the system dynamic model is calculated, as well as the global control signals (vector $\tilde{\tau}$ shown in Figs. 3.1 and 3.2) at the tactical control level. The P2 processor, ought also to carry out real-time "distribution" of the control signals onto the particular actuators at the executive control level. After that, input signals

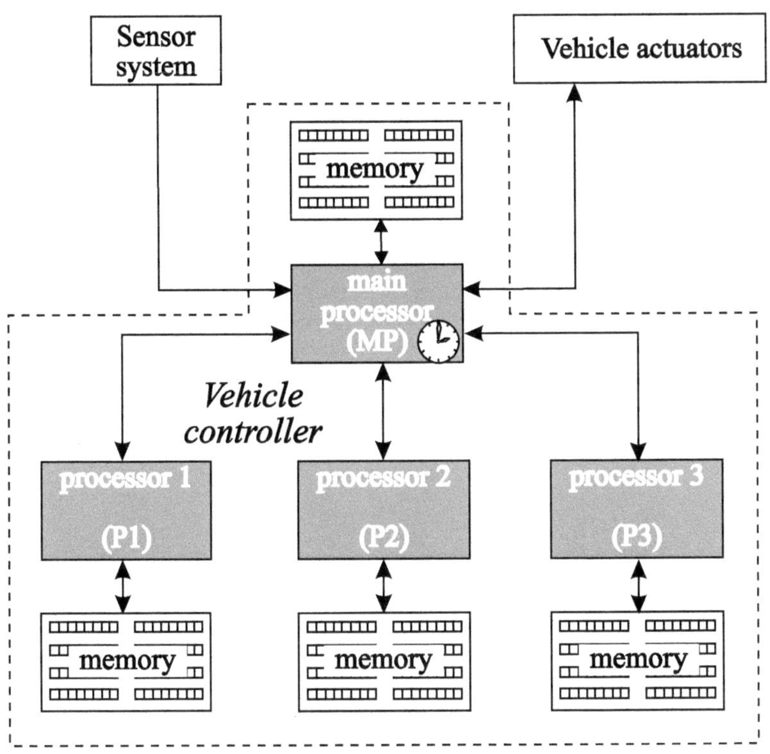

Figure 3.21: Architecture of the multiprocessor dynamic controller for integrated active control of road vehicles

for particular actuators can be determined. In the case the performing of this function should be critical in real-time, it would be necessary to install one additional processor. Its task would be to calculate the unknown values of reference signals, representing input signals to the executive level. The P3 processor in the frame of the proposed architecture would calculate the local control signals for the existing vehicle actuators. This function of the P3 processor may be dislocated to a larger number of processors which would serve the particular groups of actuators, or each actuator individually. Each of the mentioned processors (P1, P2, and P3) has its own memory for current saving of data during the operation.

The proposed dynamic controller architecture does not take into account calculation of the feedforward control signals at the outputs of the neuro-compensator. Due to the fact that the neuro-compensator is composed of a multilayered NN with a relatively large number of perceptrons in its hidden layers, the multiplication of a higher rank matrix has to be performed to calculate the NN output signals. For this operation one more supplementary processor has to be added to the existing multiprocessor structure.

Bibliography

[1] Bender, J., G., "An Overview of Systems Studies of Automated Highway Systems", *IEEE Transactions on Vehicular Technology*, Volume 40, No. 1, pp. 82-99, 1991

[2] Fenton, R., E., Mayhan, R., J., "Automated Highway Studies at The Ohio State University-An Overview", *IEEE Transactions on Vehicular Technology*, Volume 40, No. 1, pp. 100-113, 1991

[3] Shladover, S., E., Desoer, Ch., A., Hedrick, J., K., Tomizuka, M., Walrand, J., Zhang, W., McMahon, D., H., Peng, H., Sheikholeslam, S., McKeown, N., "Automatic Vehicle Control Developments in the PATH Program", *IEEE Transactions on Vehicular Technology*, Volume 40, No. 1, pp. 114-130, 1991

[4] Peng, H., Tomizuka, M., "Program on Advanced Technology (PATH) Program - Lateral Control of Front-Wheel-Steering Rubber-Tire Vehicles", Institute of Transportation Studies, University of California at Berkeley, *Technical report UCB-ITS-PRR-90-5*, UC Berkeley, July, 1990

[5] Hedrick, J., K., Tomizuka, M., Varaiya, P., "Control Issues in Automated Highwat Systems", *IEEE Control Systems Magazine*, Vol. 14, No. 6, pp. 21-32, 1994

[6] O'Brien, R., T., Iglesias, P., A., Urban, T., J., "Vehicle Lateral Control for Automated Highway Systems", *IEEE Transactions on Control Systems Technology*, Vol. 4, No. 3, pp. 266-273, May, 1996

[7] Vukobratović. M., Stojić, R., Modern Aircraft Flight Control, *Berlin: Springer-Verlag*, 1988

[8] Vukobratović, M., Stokić, D., Kirćanski, N., Non-Adaptive and Adaptive Control of Manipulation Robots, *Berlin-Heidelberg-New York-Tokyo: Springer Verlag*, 1985

[9] Irmscher, S., Hees, E., Kutsche, T., "A Controlled Suspension System with Continuously Adjustable Damping Force", *Proceedings of International Symposium on Advanced Vehicle Control for Active Safety and Ride Comfort-AVEC '94*, pp. 325-330, Tsukuba, Japan, 1994

[10] Yi, K., Oh, T., Suh, M., W., "A Robust Semi-active Suspension Control Law to Improve Ride Quality", *Proceedings of International Symposium on Advanced Vehicle Control for Active Safety and Ride Comfort-AVEC '94*, pp. 195-199, Tsukuba, Japan, 1994

[11] Kimbrough, S., "Bilinear modeling and Regulator of Variable Component Suspensions", *ASME WAM, AMD*, vol. 80, pp. 115-133, 1986.

[12] Pham, H., Hedrick, K., Tomizuka, M., "Autonomous Steering and Cruise Control of Automobiles via Sliding Mode Control", *Proceedings of International Symposium on Advanced Vehicle Control for Active Safety and Ride Comfort-AVEC '94*, pp. 444-448, Tsukuba, Japan, 1994

[13] Vallurupalli, S., S., Dukkipati, R., V., Osman, M., O., M., "Adaptive Active Suspension for a Half Car Model with a Stochastic Dirt Road Input" *Proceedings of International Symposium on Advanced Vehicle Control for Active Safety and Ride Comfort-AVEC '94*, pp. 367-371, Tsukuba, Japan, 1994

[14] Kacharoo, P., Tomizuka, M., "Vehicle Control for Automated Highway Systems for Improved Lateral Maneuverability", *Proc. of IEEE Int. Conference on Systems, Man and Cybernetics*, Vol. 1, pp. 777-781, Vancouver, Canada, 1995

[15] Ono, E., Yamashita, M., Yamauchi, Y., "Distributed Hierarchy Control of Active Suspension System", *Proceedings of International Symposium on Advanced Vehicle Control for Active Safety and Ride Comfort-AVEC '94*, pp. 373-378, Tsukuba, Japan, 1994

[16] Rodić, A., D., Vukobratović, M., K., "Contribution to the Integrated Control Synthesis of Road Vehicles", *IEEE Transactions on Control Systems Technology*, Vol. 7, No. 1, pp. 64-78, 1999

[17] Peng, H., Tomizuka, M., "Lateral Control of Front-Wheel-Steering Rubber-Tire Vehicles", *Report UCB-ITS-PRR-90-5 of PATH Program*, Institute of Transportation Studies, University of California at Berkeley, USA, 1990

[18] Pasterkamp, W., R., Pacejka, H., B., "The Tyre as a Sensor to Estimate Friction", *Proceedings of International Sysmposium on Advanced Vehicle Control-AVEC'94*, pp. 521-526, Tsukuba, Japan, October, 1994

[19] Schlüter, F., Wältermann, P., "Hierachical Control Structures for Hybrid Vehicles - modeling, Simulation, and Optimization", *Preprints of the First IFAC - Workshop on Advances in Automotive Control*, Monte-Verita, Ascona, Switzerland, March, 1995

[20] Nametz, J., Smith, R., Sigman, D., "The Design and Testing of a Microprocessor Controlled Four Wheel Steer Concept Car", *SAE paper*, No. 885087

[21] Merrit, H., Hydraulic Control Systems, *John Willey & Sons*, 1967

[22] Gibson, J., E., Tuteur, F., B., Control System Components, *Mc Graw-Hill Book Company Inc.*, New York, 1958

[23] Isermann, R., "Results on the Simplification of Dynamic Process Models", *International Journal of Control*, pp. 149-159, 1973

[24] Chiu, K., C., Smith, C., L., "Digital Control Algorithms, Part III, Tuning PI and PID controllers", *Instrum. Contr. Syst.*, pp. 41-43, December 1973

[25] Isermann, R., Digital Control Systems, *Springer-Verlag*, Berlin, 1981

[26] Sakai, H., "Theoretical and Experimental Studies on the Dynamical Properties of Tires, Part 1: Review of Theories of Rubber Friction", *International Journal of Vehicle Design*, Vol. 2, No. 1, pp. 78-110, 1981

[27] Sakai, H., "Theoretical and Experimental Studies on the Dynamical Properties of Tires, Part 4: Investigations of the Influences of Running Conditions by Calculation and Experiment", *International Journal of Vehicle Design*, Vol. 3, No. 3, pp. 333-375, 1982

[28] Bakker, E., Pacejka, H., B., Lidner, L., "A New Tire Model with an Application in Vehicle Dynamics Studies", *SAE Paper No. 890087*, 1989

[29] Pacejka, H., B., Bakker, E., "The Magic Formula Tyre Model", *Vehicle System Dynamics*, Vol. 21, ISBN 90-265-1332-1, 1993

[30] Narendra, K., S., Parthasarathy, K., "Identification and Control of Dynamical Systems Using Neural Networks", *IEEE Transactions on Neural Networks*, Vol. 1, No. 1, pp. 4-27, March 1990

[31] Tao, J., M., "Application of Neural Network with Real-Time Training to Robust Position/Force Control", *Proceedings of the IEEE International Conference on Robotics and Automation*, Atlanta, USA, pp. 142-148, May 1993

[32] Moran, A., Hasegawa, T., Nagai, M., "Continuously Controlled Semi-Active Suspension Using Neural Networks", *Proceedings of International Symposium on Advanced Vehicle Control for Active Safety and Ride Comfort-AVEC'94*, pp. 305-310, Tsukuba, Japan, October, 1994

[33] Nagai, M., Ueda, E., Moran, A., "Integration of Linear Systems and Neural Networks for the Design of Nonlinear Four Wheel Steering Systems", *Proceedings of International Symposium on Advanced Vehicle Control for Active Safety and Ride Comfort-AVEC'94*, pp. 153-158, Tsukuba, Japan, October, 1994

[34] Ozaki, T., Suzuki, T., Furuhashi, T., Okuma, S., Uchikawa, Y., "Trajectory Control of Robotic Manipulators Using Neural Networks", *IEEE Transactions on Industrial Electronics*, Vol. 38, No. 3, pp. 195-202, June, 1991

[35] Katić, D., Vukobratović, M., "Connectionest Approaches to Control of Manipulation Robots at the Executive Hierarchical Level: An Overview", *Journal of Intelligent and Robotic Systems*, Vol. 10, pp. 1-36, 1994

[36] Wasserman, P., D., Neural Computing - Theory and Practice, *New York: Van Nostrand Reinhold*, 1989

[37] Kuschewski, J., G., Hui, S., Zak, S., H., "Application of Feedforward Neural Network to Dynamical System Identification and Control", *IEEE Transactions on Control Systems Technology*, Vol. 1, No. 1, pp. 37-49, March 1993

[38] Demuth, H., Beale, M., "Neural Network Toolbox for use with MATLAB", *The Mathworks Inc.*, August 1992

Chapter 4

Stability Conditions of Road Vehicles

Abstract

This chapter describes the conditions for practical stability of road vehicle in the interaction with its dynamic environment. Stability analysis is based on the knowledge of the model of road vehicle dynamics and the model of its dynamic environment developed specially for that purpose and for the purpose of the synthesis of dynamic control laws applied in the system of automated control. Stability conditions for two dynamic control laws, pure position control and combined position-force control, were derived. In the presented example, the procedure for determining control parameters was described using the vehicle practical stability test. Besides, the characteristic simulation results, obtained by applying the proposed dynamic position-force control law, were presented for the case of a stable closed-loop system with variable system parameters. Finally, fundamental characteristics of the novel approach are pointed out and the possible research directions are outlined.

4.1 Introductory Aspects and Research Objectives

It can be stated that in the available literature the problem of road vehicle stability has not been treated adequately to its significance. For an automated vehicle as a technical system, stable behavior assumes that in any time instant the system maintains its current position and velocity within the allowed range of deviation around their nominal states. The objective of this chapter is to show the way in which a stable vehicle behavior can be guaranteed simultaneously in the main six directions of motion under the conditions of the influence of certain external perturbations and uncertainties of parameters, existing in the considered system.

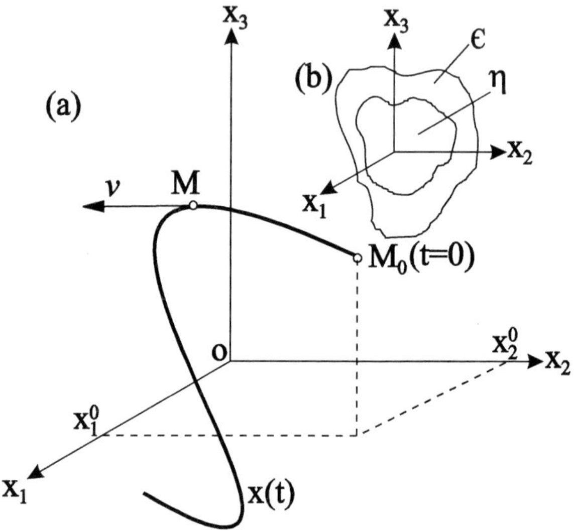

Figure 4.1: Illustration of Lyapunov's stability [2]: a) system trajectory x(t) in the state space, b) regions in the state space important for the system's stability

Undisturbed motion (stationary condition) of the system is stable according to Lyapunov [1, 2] if in a given arbitrarily small region ϵ (Fig. 4.1b) one can determine the region η such that under the initial conditions within the region η, a state trajectory x(t), i.e. a disturbed motion, does not go outside the area ϵ in the state space after an arbitrarily long period of time t from the instant of disturbance of the initial conditions x$_1$(0), x$_2$(0), x$_3$(0) (Fig. 4.1a) [2].

The notion of Lyapunov's stability can also be formulated analytically. Undisturbed motion exhibits Lyapunov's stability if at an arbitrarily small positive number ϵ is possible to find a positive number η (dependent of ϵ), such that under the initial conditions

$$|x_i^0| < \eta(\epsilon) \quad (i = 1, 2, \ldots, n) \tag{4.1}$$

the solution of the differential state equations satisfies the condition:

$$|x_i(t)| < \epsilon \quad (i = 1, 2, \ldots, n) \tag{4.2}$$

where x$_i$ is the i-th state variable in the state vector x.

If it is assumed that an arbitrary nonlinear system is asymptotically stable in the given region of initial conditions η (Fig. 4.1b), then all the system's trajectories x(t) whose origins are within the region η around the coordinate frame origin (Fig. 4.1b) will asymptotically converge with time to the coordinate origin.

By the definition of Lyapunov's stability, only a stable closed-loop system returns to its equilibrium state, but the time needed for that is not defined. Because of that, a task of crucial importance to the control designer is to synthesize a most suitable control structure and determine such system parameters that are capable of ensuring not only stability of the system equilibrium but also its so-called practical stability[1], as well as the characteristic features of the transient regime. With the majority of known technical systems it is more justifiable in the engineering sense to examine the system's practical stability. In practice, there are many technical systems which are unstable when observed in a relatively short time interval but despite of that they are asymptotically stable, for example, in an infinite time interval. For this reason, in technical practice it is more justifiable to examine the system's practical stability than its asymptotic (exponential) Lyapunov's stability.

One of the main problems in the synthesis of control laws represents the uncertainties of the dynamic models of the road vehicle and its environment. The uncertainties in the dynamic model of the environment can have especially high influence because of the difficulties in identification/prediction of the environment parameters and behavior of the environment itself. Therefore, it is of major importance to test the robustness of the synthesized control laws with respect to these model uncertainties. In this chapter we describe a method for an accurate and effective analysis of the influence of these uncertainties upon the system performance. To make such an analysis possible, the conditions for practical stability of the road vehicle with respect to the programmed motion and forces of interaction with the environment have been established for the first time. Taking into account the uncertainties of the model parameters, as well as the diverse external perturbations which may not expire with time (although being constrained), it may be difficult to achieve asymptotic (exponential) stability of the system. Therefore, it is of practical interest to demand more relaxed stability conditions, i.e. to consider the so-called practical stability of the system. Practical stability of a road vehicle is defined by specifying the finite regions around the desired position and force trajectories within which the vehicle's actual position coordinates and forces have to be during the motion along the given path. It is assumed that the inaccuracies of the model parameters (of both the vehicle and environment) are constrained.

The conditions of practical stability of the road vehicle interacting with the dynamic environment derived in this chapter enabled us to study the issue of the model uncertainties without any approximation, i.e. a correct examination of the influence of these uncertainties upon the different control laws. Also, we explain here a procedure enabling stability analysis of the considered closed-loop system, as well as the procedure for determining unknown control gains in the applied control algorithm. The same procedure can be suitable for the use in

---

[1]Under this notion we understand the system's stability in a limited time interval and within the imposed stability regions, which will be explained in detail in the subsequent sections of this chapter. Also, such stability has been termed as "technical stability".

the system design, too.

Up to now, simultaneous stability of the vehicle in the main six directions of motion, i.e. longitudinal, lateral and vertical[2], as well as the rolling, pitching, and yawing directions around the corresponding axes of inertia has not been considered. All that has been done in this domain can be mainly reduced to the examination of the vehicle stability in the state space or in the phase plane for a particular direction of motion, as in [3]. For this purpose, a simple planar vehicle model with three DOFs has been used. Bearing in mind that a road vehicle is a spatial large-scale dynamic system, the occurrence of dynamic coupling between certain system DOFs cannot be ignored. Because of that some other approach to stability analysis should be applied if we want to be theoretically consistent and practically valid. This assumes that the system's stability analysis could be strictly carried out on the basis of a more complex model of the vehicle dynamics. Thus, the conditions for a simultaneous practical stability of the road vehicle in the six main directions of motion will be given for the first time in this research monograph. A similar procedure has been developed earlier for some other dynamic systems such as manipulation robots [4, 5], aircrafts [6], etc. The basis for stability analysis of the considered system makes a spatial, nonlinear model of the vehicle dynamics presented in Chapter 2. It has been developed for that purpose and for the need of synthesizing the appropriate control laws. The research described in this chapter is based on the road vehicle model defined in the state space.

4.2   Definition of Control Task

The task of automated control of road vehicles is to ensure stability, ride quality, and accuracy of tracking of the desired motion along the nominal trajectories in the six main coordinate directions. If the vector of the current vehicle position in the time instant t is denoted by $q(t)$, then the nominal vehicle trajectory can be described by the corresponding (6×1) vector $q_0(t)$ and the corresponding nominal velocity vector $\dot{q}_0(t)$. The elements of the nominal vectors of position $q_0(t)$ and velocity $\dot{q}_0(t)$ in the six considered directions are:

$$q_0 = [x_0 \ y_0 \ \varepsilon_0 \ z_0 \ \Phi_0 \ \theta_0]^T, \quad \dot{q}_0 = [\frac{dx_0}{dt} \ \frac{dy_0}{dt} \ \frac{d\varepsilon_0}{dt} \ \frac{dz_0}{dt} \ \frac{d\Phi_0}{dt} \ \frac{d\theta_0}{dt}]^T$$

If the nominal vehicle trajectory $(q_0(t), \dot{q}_0(t))$ of the vehicle is known, then the corresponding nominal value of the force-moment vector $F_0(t)$ can be determined by using the environment model (2.73). By means of this vector we can describe the reaction of the dynamic environment upon the vehicle body. These force-moment components act upon the vehicle body MC during motion. The vector of nominal external forces and moments F_0 acting on the vehicle body is of the following form:

$$F_0 = [F_X^0 \ F_Y^0 \ M_Z^0 \ M_X^0 \ M_Y^0 \ F_Z^0]^T \tag{4.3}$$

---

[2]Here, the "vertical" is taken conditionally, considering the direction perpendicular to the road surface.

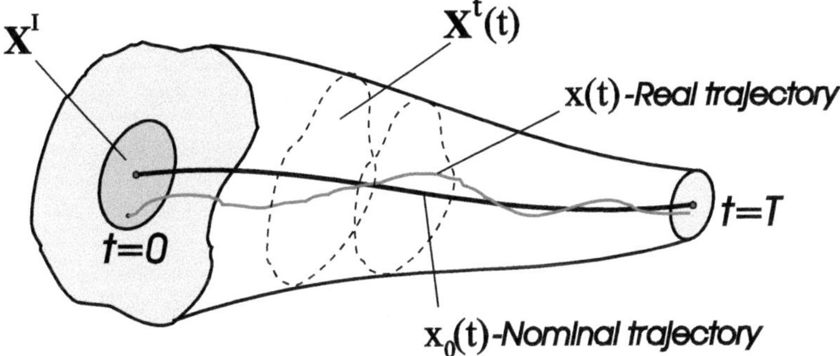

Figure 4.2: Graphical interpretation of control task - prescribed practical stability regions

The nominal system trajectories in the state space are defined as the vector:

$$x_0(t) = [q_0^T(t) \ \dot{q}_0^T(t)]^T \tag{4.4}$$

Now the system's practical stability w.r.t. the nominal state trajectory $x_0(t)$ is defined by setting the finite regions around the nominal trajectory in such a way that the real state coordinates $x(t)$ are inside the prescribed regions all the time of motion. Hereby it is understood that the values of the system and environment parameters d are variables within the allowable range. If $x_0(t)$ denotes the system nominal trajectory (4.4), then the control task can be formulated in the following way. *The control has to ensure that for $\forall x(0) \in \mathbf{X^I}$ and $\forall d \in \mathbf{D}$ it follows that $x(t) \in \mathbf{X^t}(t)$ for $\forall t \in \mathbf{T}$, where $\mathbf{X^I}$ and $\mathbf{X^t}(t)$ are the prescribed finite regions in the state space around the nominal trajectory $x_0(t)$ and, where $\mathbf{T} = (t, t \in (0, t_1))$. Here, t_1 is a predefined time interval. It is taken that $x_0(0) \in \mathbf{X^I}$, $x_0(t) \in \mathbf{X^t}(t)$, $\forall t \in \mathbf{T}$, $\mathbf{X^t}(0) \supset \mathbf{X^I}$.* This formulation of the control task can be reinterpreted in the following way. For a given desired trajectory $x_0(t)$ an initial deviation from the nominal trajectory is allowed in such a way that the state vector $x(0)$ has to belong to the predefined region $\mathbf{X^I}$ around $x_0(0)$. The control task has to ensure the system's motion in the state space $x(t)$ is tracking the desired trajectory $x_0(t)$ with the allowed constrained error $\Delta x(t)$. Such a requirement that the system state trajectory $x(t)$ belongs to a predefined region $\mathbf{X^t}(t)$ about the nominal trajectory $x_0(t)$ must be satisfied in the time interval $\mathbf{T}$ defined in advance. Graphical interpretation of the control task is shown in Fig. 4.2. All that we have just said should be valid for each possible value d of the vehicle's model parameters and for any parameter of its dynamic environment. Taking into account the expression (2.73), describing the model of the "dynamic environment", and using the results published in [7], the realization of the defined control task guarantees also a convergent tracking of the desired force trajectories $F_0(t)$. This concretely means that for $F(0) \in \mathbf{F^I}$ and for $\forall d \in \mathbf{D}$ it follows that $F \in \mathbf{F^t}(t)$, for $\forall t \in \mathbf{T}$, where $\mathbf{F^I}$ and $\mathbf{F^t}(t)$ are the regions in the $(m \times 1)$ space around the nominal trajectory $F_0(t)$. Let us point out that the regions $\mathbf{F^I}$ and $\mathbf{F^t}(t)$ have to correspond to the regions $\mathbf{X^I}$ and $\mathbf{X^t}(t)$. Aimed at simplifying practical stability analysis, specific forms of the finite regions $\mathbf{X^I}$ and $\mathbf{X^t}(t)$ are considered: $\mathbf{X^I} = \{x(0) : \|\Delta x(0)\| < \underline{X}^I\}$, $\mathbf{X^t}(t) = \{x(t) : \|\Delta x(t)\| < \underline{X}^t exp(-\alpha t)\}$, for $\forall t \in \mathbf{T}$, where $\underline{X}^t > \underline{X}^I > 0$, $\alpha > 0$. The quantities $\underline{X}^t$, $\underline{X}^I$ and α designate real positive numbers, $\|.\|$ is the Euclidian norm of the corresponding vector, while Δx denotes the $(2n)$ vector of the state coordinates deviations from the nominal trajectory $x_0(t)$:

$$\Delta x(t) = x(t) - x_0(t) \tag{4.5}$$

4.3   General Stability Conditions

In this section general conditions of the system's practical stability for a chosen control law will be presented. According to Weissenberger [8], Michael [9], Vukobratovic and Stokic [4, 5], the system is practically stable w.r.t. ($\mathbf{X^I}$, $\mathbf{X^t}(t)$, $\mathbf{T}$, see Fig. 4.2) if there exists a real, continual, differentiable function $v(t, x)$ and a real time function $\Psi(t)$ which is integrable along a time interval $\mathbf{T}$, such that the following conditions are satisfied:

$$\dot{v}(t, x) \le \Psi, \quad \forall x \in \tilde{X}^t(t), \quad \forall t \in \mathbf{T} \tag{4.6}$$

$$\int_0^t \Psi(t') \, dt' < v_m^{\partial X^t}(t) - v_M^{\partial X^I}(0), \quad \forall t \in \mathbf{T}. \tag{4.7}$$

where: $\partial \mathbf{X}(t)$ is the boundary of the corresponding region (see Fig. 4.2) and $\tilde{X}^t(t) = \mathbf{X^t}(t) - \underline{X}^I exp(-\alpha t)$. In the equation (4.6) $\dot{v}$ denotes the time derivative of the function $v(t, x)$ along the solution of the closed-loop control system. The symbols v_m and v_M denote the minimal and the maximal value of the function $v(t, x)$ on the corresponding region boundaries. The above conditions of the system's practical stability have been proved in [9]. *The stability conditions (4.6) and (4.7) are conservative for the reason that the lack of their fulfilment does not mean that the system is unstable. They are only sufficient but not necessary conditions of stability.*

4.4   Stability Conditions for Proposed Control Laws

General conditions of vehicle's practical stability were implemented for the proposed dynamic control laws (3.3) and (3.8). In this section we will derive first the concrete expressions for the vehicle stability conditions in case of the application of the position control algorithm (3.3). After that, we will present concrete expressions for the vehicle stability conditions in the case of the application of the combined position-force dynamic control law.

4.4.1 Implementation of the Position Control Law

In order to derive the conditions of practical stability of the system involving a control law defined in advance, it is necessary to determine the model of the closed-loop system. This is done in such a way that the expressions for the corresponding control law, whose stability characteristics are to be investigated, is introduced into the vector relation (2.28). In this case it is a purely positional dynamic control law defined by the relation (3.3). Bearing in mind that the nominal system trajectory $q_0^{(2)}(t)$ is determined in an indirect way, by calculating it from the dynamic environment model (2.73), the inevitable parametric inaccuracies of the model have to be taken into account. They, to a greater or smaller extent, influence the accuracy of calculation of the nominal system trajectory. For the assumed inaccurate parameter values of the environment model $\tilde{d}$, the approximate values $\hat{q}_0^{(2)}(t)$ of the trajectories in the corresponding directions are calculated in the following way:

$$\hat{\ddot{q}}_0^{(2)} = \hat{M}_{22}^{-1}\left(-F_0^{(2)} - \hat{M}_{21}\ddot{q}_0^{(1)} - \hat{L}^{(2)}\right) \tag{4.8}$$

where $\hat{M}_{21}$, $\hat{M}_{22}$ and $\hat{L}^{(2)}$ denote the values of the matrices and vectors M_{21}, M_{22} and $L^{(2)}$ from the model (2.73) for the assumed values of the environment parameters $\tilde{d}$. In the x, y and ε directions, the trajectories are prescribed in a direct way (not by the calculation based on the model as is done for the values z, Φ and θ). As a consequence, the nominal trajectory $q_0(t)$ (calculated from the model) will deviate from the real nominal trajectory $\hat{q}_0(t)$ to the extent depending on how accurately were determined the values of the environmental parameters $\tilde{d}$. The inaccuracies of the applied model and, consequently, of the nominal trajectories, have a direct influence upon the system's stability [10]. The extent of their influence can be determined by analyzing the stability conditions, which will be explained later on.

Taking into account the numerous inevitable inaccuracies of the environment approximation model, the position control law (3.3) can be written in an extended form by introducing a supplementary term which takes into account the inaccuracies of calculation of the nominal trajectory $q_0(t)$. So, the expression (3.3) can be written as:

$$\begin{aligned}\tau = \ & J^T\{\dot{H}\left[(\ddot{q}_0 + K_V\Delta\dot{q} + K_P\Delta q) + R\,\Delta\ddot{q}_0 + R\left(K_V\Delta\dot{q}_0 + K_P\Delta q_0\right)\right] \\ & + \ \hat{h} - F\}\end{aligned} \tag{4.9}$$

where the terms $R \cdot \Delta\ddot{q}_0$ and $R \cdot (K_V\Delta\dot{q}_0 + K_P\Delta q_0)$ were introduced as the so-called "correction" terms, for the reason that the calculation of the programmed trajectory was performed with the inaccurate values of the environment model parameters. Hereby, the following notations were used:

$$\Delta q_0(t) = q_0(t) - \hat{q}_0(t), \quad \Delta\dot{q}_0(t) = \dot{q}_0(t) - \hat{\dot{q}}_0(t), \quad \Delta\ddot{q}_0(t) = \ddot{q}_0(t) - \hat{\ddot{q}}_0(t)$$

which are representing the (6×1) vectors of position and acceleration errors due to the calculation of nominal trajectory based on inaccurate parameter values of the environment model. The matrix R represents the $(n \times n)$ "uncompleted" inertia matrix H which is taking into account the previously described model decoupling. This matrix possesses a zero submatrix in the position that corresponds to the directions in which nominal trajectory was explicitly prescribed and a non-zero submatrix in the positions corresponding to the directions in which nominal trajectory was implicitly prescribed by using the environment model. The matrix R is of the following form:

$$\hat{H} = [\hat{H}_{1_{sub}}^{n \times n_1} \ \hat{H}_{2_{sub}}^{n \times (n-n_1)}]; \quad \hat{H}_2 = [0^{n \times n_1} \ \hat{H}_{2_{sub}}^{n \times (n-n_1)}];$$
$$R = \hat{H}^{-1} \cdot \hat{H}_2; \tag{4.10}$$

where the $(n \times n)$ matrix $\hat{H}_2$ contains the corresponding submatrix elements of $\hat{H}$. In the way previously described, the matrix R ensures that the inaccuracies of the dynamic environment parameters are taken into account only in the directions in which the nominal force trajectory $F_0^{(2)}(t)$ has been prescribed explicitly. In the other directions (x, y and ε), the vehicle's nominal trajectory is "exact" ($\hat{q}_0^{(1)}(t) \equiv q_0^{(1)}(t)$), because this vector was prescribed in advance by giving the values in a direct way.

Using the vehicle model (2.28) and the control law (3.3), the following relation is obtained:

$$H\ddot{q} + h = J^{-T}\hat{J}^T\{\hat{H}\,[(\ddot{q}_0 + K_V \Delta \dot{q} + K_P \Delta q) + R\,\Delta \ddot{q}_0 +$$
$$+ R\,(K_V \Delta \dot{q}_0 + K_P \Delta q_0)] + \hat{h} - F\} + F \tag{4.11}$$

After certain mathematical transformations it can be reduced to the form:

$$\Delta \ddot{q} = (Q - I)\ddot{q}_0 + Q(K_V \Delta \dot{q} + K_P \Delta q) + H^{-1}(J^{-T}\hat{J}^T\hat{h} - h) +$$
$$+ H^{-1}(I - J^{-T}\hat{J}^T)F + Q\,R\Delta \ddot{q}_0 + Q\,R(K_V \Delta \dot{q}_0 + K_P \Delta q_0) \tag{4.12}$$

The matrix Q in the expression (4.12) is of the form:

$$Q = H^{-1}J^{-T}\hat{J}^T\hat{H} \tag{4.13}$$

Having in mind that the state vector was defined in a way described by the expression (2.28), the equation (4.12) can be written in the form of a closed-loop system model in the state space. The obtained model of the system's deviation from the nominal trajectory $x_0(t)$ can be expressed in a condensed form as:

$$\Delta \dot{x} = \Delta f(\Delta x, \hat{x}_0, d) + \Delta G(\Delta x, d)F \tag{4.14}$$

where $\Delta f(\Delta x, \hat{x}_0, d)$ is a $(2n \times 1)$ vector, and $\Delta G(\Delta x, d)$ is a $(2n \times n)$ matrix. They are defined by the following expressions:

$$\Delta f(\Delta x, \hat{x}_0, d) = f_1 + f_2 + f_3 + f_4 + f_5;$$
$$f_1 = A_1 \Delta x = \begin{bmatrix} 0 & I \\ QK_P & QK_V \end{bmatrix} \Delta x;$$

$$f_2 = A_2 \dot{x}_0 = \begin{bmatrix} 0 & 0 \\ 0 & H^{-1}(J^{-T}\hat{J}^T\hat{H} - H) \end{bmatrix} \dot{x}_0;$$

$$f_3 = \begin{bmatrix} 0 \\ H^{-1}(J^{-T}\hat{J}^T\hat{h} - h) \end{bmatrix};$$

$$f_4 = A_4(\dot{x}_0 - \hat{\dot{x}}_0) = \begin{bmatrix} 0 & 0 \\ 0 & -Q\,R \end{bmatrix} (\dot{x}_0 - \hat{\dot{x}}_0);$$

$$f_5 = A_5\,\Delta x_0 = \begin{bmatrix} 0 & 0 \\ -QRK_P & -QRK_V \end{bmatrix} \Delta x_0;$$

$$\Delta G(\Delta x, d) = \begin{bmatrix} 0 \\ H^{-1}(I - J^{-T}\hat{J}^T) \end{bmatrix}; \tag{4.15}$$

The matrices A_1, A_2, A_4 and A_5 are $(2n \times 2n)$ matrices, Q is an $(n \times n)$ matrix given by the relation (4.13), Δx_0 is a $(2n \times 1)$ difference vector of the nominal trajectory $\hat{x}_0(t)$ based on the inaccurate values of the model parameters $\tilde{d}$ from the "accurate" nominal trajectory $x_0(t)$ that corresponds to the quantities $q_0^{(1)}(t)$ and $F_0^{(2)}(t)$, i.e.:

$$\Delta x_0(t) = x_0(t) - \hat{x}_0(t) = [0^{1 \times n_1} \ \Delta q_0^{(2)\,T} \ 0^{1 \times n_1} \ \Delta \dot{q}_0^{(2)\,T}]^T \tag{4.16}$$

A simple interpretation of the feedback loop terms in the expression (4.15) would be as follows. The term f_1 represents the "feedback" term that directly depends on the influence of the inaccuracies of the vehicle dynamic model parameters, i.e. it depends directly on the inaccuracies of calculation of the matrices J and H. The term f_2 depends also on the inaccuracies of the applied model (i.e. the expressions for J and H), as well as on the prescribed nominal values $\dot{x}_0(t)$ of velocity and acceleration. Only in the case when the real values of the system parameters d are equal to the assumed ones, d_0, i.e. when the nominal accelerations $\dot{x}_0 \equiv 0$, then the term $f_2 \equiv 0$. The term f_3 depends on the vehicle's model inaccuracies, too. These inaccuracies are caused by the calculation error of the gravitational and centrifugal forces h (if $d \equiv d_0$ then $f_3 \equiv 0$). The term f_4 depends on the uncertainties in the environment model (i.e. on the inaccuracy of the expressions for M and L in the relation (2.73)), causing inaccurate calculation of the nominal trajectory $\ddot{q}_0^{(2)}(t)$ from (4.8). In the case when $\tilde{d} \equiv \tilde{d}_0$, then $\hat{\dot{q}}_0^{(2)}(t) \equiv \dot{q}_0^{(2)}(t)$ and $f_4 \equiv 0$. The term f_5 is also a consequence of the inaccurate calculation of the desired trajectory, causing an error in the feedback loop. If $\tilde{d} \equiv \tilde{d}_0$ then $\hat{q}_0^{(2)} \equiv q_0^{(2)}$ and $\Delta x_0 \equiv 0$, so that $f_5 \equiv 0$, too. The expression $\Delta G \cdot F$ describes the influence of the force $F(t)$ upon the system. Values of the expressions depend on the inaccuracy of calculation of the Jacobian matrix J. If $d \equiv d_0$, then also $J \equiv \hat{J}$ and $\Delta G \equiv 0$.

4.4.1.1 Derivation of Terms of Stability Conditions

Let us choose the function $v(t, \Delta x)$ in the form:

$$v(t, \Delta x) = (\Delta x^T H \Delta x)^{\frac{1}{2}} \tag{4.17}$$

where H is a $(2n \times 2n)$ positive definite symmetric matrix[3], selected in such a way to satisfy the following inequality of Lyapunov's type:

$$HA_1 + A_1^T H \leq -2\,min|\sigma(A_1)| \cdot H \tag{4.18}$$

The derivative of the function v along the solution of the equation (4.14) can be written in the form:

$$
\begin{aligned}
\dot{\mathrm{v}} \;=\;& \frac{1}{2\,\mathrm{v}}[\Delta \mathrm{x}^T(HA_1 + A_1^T H)\Delta \mathrm{x}] + \\
+\;& (grad(\mathrm{v}))^T[A_2\dot{\mathrm{x}}_0 + f_3 + A_4\Delta\dot{\mathrm{x}}_0 + A_5\Delta \mathrm{x}_0] + \\
+\;& (grad(\mathrm{v}))^T \Delta G(\Delta \mathrm{x})F
\end{aligned} \tag{4.19}
$$

The gradient of the vector function v has the following elements:

$$(grad(\mathrm{v}))^T \;=\; \left[\; \frac{\partial \mathrm{v}}{\partial \Delta \mathrm{x}_1} \quad \frac{\partial \mathrm{v}}{\partial \Delta \mathrm{x}_2} \quad \cdots \quad \frac{\partial \mathrm{v}}{\partial \Delta \mathrm{x}_{2n-1}} \quad \frac{\partial \mathrm{v}}{\partial \Delta \mathrm{x}_{2n}} \;\right] \tag{4.20}$$

$$
\begin{aligned}
\frac{\partial \mathrm{v}}{\partial \Delta \mathrm{x}_1} = \frac{1}{2\cdot \mathrm{v}} \;&\cdot\; (2\cdot H_{1,1}\Delta \mathrm{x}_1 + H_{2,1}\Delta \mathrm{x}_2 + \cdots + H_{2n,1}\Delta \mathrm{x}_{2n} + \\
&+\; H_{1,2}\Delta \mathrm{x}_2 + H_{1,3}\Delta \mathrm{x}_3 + \cdots + H_{1,2n}\Delta \mathrm{x}_{2n}) \\[4pt]
\frac{\partial \mathrm{v}}{\partial \Delta \mathrm{x}_2} = \frac{1}{2\cdot \mathrm{v}} \;&\cdot\; (H_{1,2}\Delta \mathrm{x}_1 + 2\cdot H_{2,2}\Delta \mathrm{x}_2 + \cdots + H_{2n,2}\Delta \mathrm{x}_{2n} + \\
&+\; H_{2,1}\Delta \mathrm{x}_1 + H_{2,3}\Delta \mathrm{x}_3 + \cdots + H_{2,2n}\Delta \mathrm{x}_{2n}) \\
&\qquad \cdots \\
&\qquad \cdots \\
\frac{\partial \mathrm{v}}{\partial \Delta \mathrm{x}_{2n}} = \frac{1}{2\cdot \mathrm{v}} \;&\cdot\; (H_{1,2n}\Delta \mathrm{x}_1 + H_{2,2n}\Delta \mathrm{x}_2 + \cdots + 2\cdot H_{2n,2n}\Delta \mathrm{x}_{2n} + \\
&+\; H_{2n,1}\Delta \mathrm{x}_1 + H_{2n,2}\Delta \mathrm{x}_2 + \cdots + H_{2n,2n-1}\Delta \mathrm{x}_{2n-1})
\end{aligned}
$$

Now the expressions for estimations of the functions given on the right-hand side of the equation (4.19) can be derived as:

$$(\mathrm{I}) \qquad \frac{1}{2\,\mathrm{v}}[\Delta \mathrm{x}^T(HA_1 + A_1^T H)\Delta \mathrm{x}] < -\min_{i,d\in \mathbf{D}} |\sigma_i(A_1)|\,\lambda_m^{1/2}(H)\,\|\Delta \mathrm{x}\|$$

$$(\mathrm{II}) \qquad (grad(\mathrm{v}))^T A_2\dot{\mathrm{x}}_0 < \sum_{j=1}^{2n} \xi_j^{(1)}|\Delta \mathrm{x}_j(t)| + \sum_{j=1}^{2n} \xi_j^{(2)}|\dot{\mathrm{x}}_{0_j}(t)| + \xi^{(3)}$$

$$(\mathrm{III}) \qquad (grad(\mathrm{v}))^T f_3 < \xi^{(4)} + \sum_{j=1}^{2n} \xi_j^{(5)}|\Delta \mathrm{x}_j(t)|$$

$$(\mathrm{IV}) \qquad (grad(\mathrm{v}))^T A_4(\dot{\mathrm{x}}_0 - \hat{\dot{\mathrm{x}}}_0) < \sum_{j=1}^{2n_1} \xi_j^{(6)}|\dot{\mathrm{x}}_{0_j}^{(1)}(t)| + \sum_{j=1}^{m_2} \xi_j^{(7)}|F_{0_j}^{(2)}| +$$

---

[3]This matrix H differs from the previously described $(n \times n)$ matrix of inertia H used in the model (2.28).

$$+ \sum_{j=1}^{2n} \xi_j^{(8)} |\Delta x_j(t)| + \xi^{(9)}$$

$$(V) \qquad (grad(v))^T A_5 \Delta \hat{x}_0 < \sum_{j=1}^{2n_1} \xi_j^{(10)} |x_{0_j}^{(1)}| + \sum_{j=1}^{m_2} \xi_j^{(11)} |F_{0_j}^{(2)}| +$$

$$+ \sum_{j=1}^{m_2} \xi_j^{(12)} \int_0^t |F_{0_j}^{(2)}| dt + \sum_{j=1}^{2n} \xi_j^{(13)} |\Delta x_j(t)| + \xi^{(14)}$$

$$(VI) \qquad (grad(v))^T \Delta G(\Delta x) F < \sum_{j=1}^{m} \xi_j^{(15)} |F_j(t)| +$$

$$+ \sum_{j=1}^{2n} \xi_j^{(16)} |\Delta x_j(t)| \qquad\qquad (4.21)$$

where: σ_i, $i = 1, 2, \ldots, 2n$, are the eigenvalues of the matrix A_1 for the assumed parameters d; λ_m is the minimal eigenvalue of the positive definite matrix H; γ is a scalar value calculated from the expression $\gamma = min|\sigma_i| \cdot \lambda_m^{1/2}(H)$; $\xi_j^{(k)}$ are positive or negative numbers in the expressions for estimation, where $k = 1, \cdots, 16$; x_j denotes the j-th component of the state vector x. The inequalities (4.21) should be valid for $\forall x \in \mathbf{X}^t(t)$, $\forall t \in \mathbf{T}$ and $d \in \mathbf{D}$. Let us notice that the inequalities denoted by (IV) and (V) in (4.21) are consequences of the calculation of the nominal trajectory $\hat{q}_0^{(2)}$ based on the "inaccurate" inverse environment model (4.8).

Substituting the expressions from (4.21) into the expression for the function derivative $\dot{v}$ (4.19), a concrete form of the function $\Psi(t)$ is obtained as:

$$\begin{aligned}
\Psi(t) \;=\; & -\gamma \, \underline{X}^I exp(-\alpha t) + \\
& + \; (\xi^{(1)} + \xi^{(5)} + \xi^{(8)} + \xi^{(13)} + \xi^{(16)}) \underline{X}^t exp(-\alpha t) + \\
& + \; (\xi^{(3)} + \xi^{(4)} + \xi^{(9)} + \xi^{(14)}) + \sum_{j=1}^{2n_1} \xi_j^{(10)} |x_{0_j}^{(1)}(t)| + \sum_{j=1}^{2n_1} \xi_j^{(6)} |\dot{x}_{0_j}^{(1)}(t)| + \\
& + \; \sum_{j=1}^{2n} \xi_j^{(2)} |\dot{x}_{0_j}(t)| + \sum_{j=1}^{m_2} (\xi_j^{(7)} + \xi_j^{(11)}) |F_{0_j}^{(2)}(t)| + \\
& + \; \sum_{j=1}^{m_2} \xi_j^{(12)} \int_0^t |F_{0_j}^{(2)}(t)| dt + \\
& + \; \sum_{j=1}^{m} \xi_j^{(15)} |F_{0_j}(t) + \underline{F}_j^t exp(-\beta t)| \qquad\qquad (4.22)
\end{aligned}$$

where $\xi^{(1)}, \xi^{(5)}, \xi^{(8)}, \xi^{(13)}$, and $\xi^{(16)}$ are the maximal values of the corresponding coefficients $\xi_j^{(1)}, \xi_j^{(5)}, \xi_j^{(8)}, \xi_j^{(13)}$ and $\xi_j^{(16)}$. $\underline{F}_j^t$, $j = 1, \ldots, m$ are positive numbers corresponding to the region $F^t(t)$. Taking into account that:

$$v_m^{\frac{\partial X^t}{}}(t) \; > \; \lambda_m^{1/2}(H) \underline{X}^t exp(-\alpha t);$$

$$v_M^{\partial X^I}(0) \quad < \quad \lambda_M^{1/2}(H)\underline{X}^I \tag{4.23}$$

and where λ_M is the maximal eigenvalue of the matrix H, the general condition for practical stability (4.7) can be written as:

$$
\begin{aligned}
\int_0^t \Psi(t')dt' \;=\; & [-\gamma\,\underline{X}^I + (\xi^{(1)} + \xi^{(5)} + \xi^{(8)} + \xi^{(13)} + \xi^{(16)})\underline{X}^t] \cdot \\[4pt]
& \cdot\;(\frac{1 - exp(-\alpha t)}{\alpha}) + \\[4pt]
+\; & (\xi^{(3)} + \xi^{(4)} + \xi^{(9)} + \xi^{(14)})t + \sum_{j=1}^{2n_1} \xi_j^{(10)} \int_0^t |\mathrm{x}_{0_j}^{(1)}(t)|dt + \\[4pt]
+\; & \sum_{j=1}^{2n_1} \xi_j^{(6)} \int_0^t |\dot{\mathrm{x}}_{0_j}^{(1)}(t)|dt + \sum_{j=1}^{2n} \xi_j^{(2)} \int_0^t |\dot{\mathrm{x}}_{0_j}(t)|dt + \\[4pt]
+\; & \sum_{j=1}^{m_2} (\xi_j^{(7)} + \xi_j^{(11)}) \int_0^t |F_{0_j}^{(2)}(t)|dt + \\[4pt]
+\; & \sum_{j=1}^{m_2} \xi_j^{(12)} \int_0^t \int_0^t |F_{0_j}^{(2)}(t)|dt^2 + \\[4pt]
+\; & \sum_{j=1}^{m} \xi_j^{(15)} (\int_0^t |F_{0_j}(t)|dt + \underline{F}_j^t \frac{1 - exp(-\beta t)}{\beta}) < \\[4pt]
<\; & \lambda_m^{1/2}(H)\underline{X}^t exp(-\alpha t) - \lambda_M^{1/2}(H)\underline{X}^I \tag{4.24}
\end{aligned}
$$

In this way, the concrete expressions for the general conditions (4.6) and (4.7) of vehicle's practical stability were determined for the case of pure positional control law.

4.4.2   Implementation of the Position-Force Control Law

In the text to follow we will derive the concrete stability conditions for the proposed combined position-force control algorithm that takes into account the overall dynamics of the road vehicle interacting with the road surface.

In order to derive practical stability conditions for the system involving the dynamic position-force control law (3.8) it is necessary to determine the model of the closed-loop system. In the case when the combined position-force control law is applied, the model of closed-loop system is obtained by combining the relation (2.28) and the control law (3.8). In this control law, for the purpose of the current consideration, the polynomial $Q(.)$ is chosen in the form of a simple P-regulator. Then, the following equation is obtained:

$$H\ddot{q} + h = J^{-T}\hat{J}^T\{\hat{H}\ddot{q}_c + \hat{h} - F\} + F \tag{4.25}$$

Expression for $\ddot{q}_c$ can be written in the form:

$$\ddot{q}_c^{(1)} \;=\; \ddot{q}_0^{(1)} + K_V^{(1)}\Delta\dot{q}^{(1)} + K_P^{(1)}\Delta q^{(1)}$$

$$\ddot{q}_c^{(2)} = -\hat{M}_{22}^{-1}\hat{M}_{21}(\ddot{q}_0^{(1)} + K_V^{(1)}\Delta\dot{q}^{(1)} + K_P^{(1)}\Delta q^{(1)}) - \hat{M}_{22}^{-1}\hat{L}^{(2)} -$$

$$- \hat{M}_{22}^{-1}F_0^{(2)} - \hat{M}_{22}^{-1}K_F^{(2)}\int_0^t (F^{(2)} - F_0^{(2)})dt$$

$$\ddot{q}_c = \begin{bmatrix} \ddot{q}_c^{(1)} \\ \ddot{q}_c^{(2)} \end{bmatrix} \tag{4.26}$$

The matrices $K_P^{(1)}$ and $K_V^{(1)}$ represent the $(n_1 \times n_1)$ matrices of the position and velocity gains. The matrix $K_F^{(2)}$ is an $(m_2 \times m_2)$ gain matrix w.r.t. force in the feedback loop.

The second expression in the set of relations (4.26) can be expanded using the relation for the environment model (2.73):

$$-\hat{M}_{22}^{-1}\hat{L}^{(2)} - \hat{M}_{22}^{-1}F_0^{(2)} = -\hat{M}_{22}^{-1}\hat{L}^{(2)} + (\hat{M}_{22}^{0\,-1} - \hat{M}_{22}^{-1})F_0^{(2)} +$$

$$+ \hat{M}_{22}^{0\,-1}(\hat{M}_{21}^0\ddot{q}_0^{(1)} + \hat{M}_{22}^0\ddot{q}_0^{(2)} + \hat{L}_0^{(2)}) \tag{4.27}$$

Also, using the environment model (2.73), the expression under the integral can be expanded further, so that:

$$\int_0^t (F^{(2)} - F_0^{(2)})dt = \int_0^t (-(\hat{M}_{21}\ddot{q}^{(1)} + \hat{M}_{22}\ddot{q}^{(2)} + \hat{L}^{(2)}) +$$

$$+ (\hat{M}_{21}^0\ddot{q}_0^{(1)} + \hat{M}_{22}^0\ddot{q}_0^{(2)} + \hat{L}_0^{(2)}))dt \tag{4.28}$$

After certain mathematical transformations, the expression (4.27) can be written as:

$$\int_0^t (F^{(2)} - F_0^{(2)})dt = -\int_0^t \hat{M}_{21}\Delta\ddot{q}^{(1)}dt - \int_0^t \hat{M}_{22}\Delta\ddot{q}^{(2)}dt -$$

$$- \int_0^t (\hat{L}^{(2)} - \hat{L}_0^{(2)})dt + \int_0^t (\hat{M}_{21}^0 - \hat{M}_{21})\ddot{q}_0^{(1)}dt +$$

$$+ \int_0^t (\hat{M}_{22}^0 - \hat{M}_{22})\ddot{q}_0^{(2)}dt \tag{4.29}$$

Taking into account the fact that the environment model inertial parameters $\hat{M}_{21}(t)$ and $\hat{M}_{22}(t)$ are generally time variable, being continuous functions over the time interval t as well as their time derivatives, then the previous expression (4.29) can be subjected to further mathematical treatment. Hereby, the rule of partial integration, given in its general form, is used for this purpose:

$$\int u(x)\, dv = u(x)\, v(x) - \int v(x)\, du \tag{4.30}$$

The expressions to appear in the further transformations of the mathematical relation (4.29) have the following forms:

$$\int_0^t \hat{M}_{21}\Delta\ddot{q}^{(1)}dt = \hat{M}_{21}\Delta\dot{q}^{(1)} - \hat{M}_{21}\Delta q^{(1)} + \int_0^t \dot{\hat{M}}_{21}\Delta q^{(1)}dt$$

$$\int_0^t \hat{M}_{22}\Delta\ddot{q}^{(2)}dt = \hat{M}_{22}\Delta\dot{q}^{(2)} - \hat{M}_{22}\Delta q^{(2)} + \int_0^t \dot{\hat{M}}_{22}\Delta q^{(2)}dt \tag{4.31}$$

Taking into account the intermediate results of integrating the products of functions (4.31), the expression (4.29) can be written in a definite form as:

$$
\begin{aligned}
\int_0^t (F^{(2)} \; - \; F_0^{(2)})dt = -\hat{M}_{21}\Delta\dot{q}^{(1)} + \dot{\hat{M}}_{21}\Delta q^{(1)} - \int_0^t \ddot{\hat{M}}_{21}\Delta q^{(1)}dt - \\
- \; \hat{M}_{22}\Delta\dot{q}^{(2)} + \dot{\hat{M}}_{22}\Delta q^{(2)} - \int_0^t \ddot{\hat{M}}_{22}\Delta q^{(2)}dt - \\
- \; \int_0^t (\hat{L}^{(2)} - \hat{L}_0^{(2)})dt + \int_0^t (\hat{M}_{21}^0 - \hat{M}_{21})\ddot{q}_0^{(1)}dt + \\
+ \; \int_0^t (\hat{M}_{22}^0 - \hat{M}_{22})\ddot{q}_0^{(2)}dt
\end{aligned}
\tag{4.32}
$$

Finally, the expression (4.26), in which the value of the "control acceleration" $\ddot{q}_c$ appearing in the control law (3.8) is determined, can be written in the form:

$$
\ddot{q}_c \;=\; \left[\begin{array}{c} \ddot{q}_c^{(1)} \\ \ddot{q}_c^{(2)} \end{array} \right] = \left[\begin{array}{c} \ddot{q}_c^{(1)} \\ \ddot{q}_c^{(I)} + \ddot{q}_c^{(II)} + \ddot{q}_c^{(III)} + \ddot{q}_c^{(IV)} + \ddot{q}_c^{(V)} + \ddot{q}_c^{(VI)} \end{array} \right]
\tag{4.33}
$$

where

$$
\begin{aligned}
\ddot{q}_c^{(I)} \;&=\; (\hat{M}_{22}^0{}^{-1}\hat{M}_{21}^0 - \hat{M}_{22}^{-1}\hat{M}_{21})\ddot{q}_0^{(1)} + \hat{M}_{22}^0{}^{-1}\hat{M}_{22}^0\ddot{q}_0^{(2)} \\[4pt]
\ddot{q}_c^{(II)} \;&=\; (\hat{M}_{22}^{-1}\hat{M}_{21}K_P^{(1)} + \hat{M}_{22}^{-1}K_F^{(2)}\dot{\hat{M}}_{21})\Delta q^{(1)} + \\
&\quad + \; (-\hat{M}_{22}^{-1}\hat{M}_{21}K_V^{(1)} + \hat{M}_{22}^{-1}K_F^{(2)}\hat{M}_{21})\Delta\dot{q}^{(1)} \\[4pt]
\ddot{q}_c^{(III)} \;&=\; \hat{M}_{22}^{-1}K_F^{(2)}\int_0^t \ddot{\hat{M}}_{21}\Delta q^{(1)}dt + \hat{M}_{22}^{-1}K_F^{(2)}\int_0^t \dot{\hat{M}}_{22}\Delta q^{(2)}dt \\[4pt]
\ddot{q}_c^{(IV)} \;&=\; -\hat{M}_{22}^{-1}K_F^{(2)}\dot{\hat{M}}_{22}\Delta q^{(2)} + \hat{M}_{22}^{-1}K_F^{(2)}\hat{M}_{22}\Delta\dot{q}^{(2)} \\[4pt]
\ddot{q}_c^{(V)} \;&=\; \hat{M}_{22}^0{}^{-1}\hat{L}_0^{(2)} - \hat{M}_{22}^{-1}\hat{L}^{(2)} + \hat{M}_{22}^{-1}K_F^{(2)})\int_0^t (\hat{L}^{(2)} - \hat{L}_0^{(2)})dt \\[4pt]
\ddot{q}_c^{(VI)} \;&=\; (\hat{M}_{22}^0{}^{-1} - \hat{M}_{22}^{-1})F_0^{(2)} - \hat{M}_{22}^{-1}K_F^{(2)}\int_0^t (\hat{M}_{21}^0 - \hat{M}_{21})\ddot{q}_0^{(1)}dt - \\
&\quad - \; \hat{M}_{22}^{-1}K_F^{(2)}\int_0^t (\hat{M}_{22}^0 - \hat{M}_{22})\ddot{q}_0^{(2)}dt
\end{aligned}
\tag{4.34}
$$

The variables M_{ij}^0, used in the previous expressions, represent the corresponding matrices in the model (2.73), calculated for the conditions of the nominal trajectory $\mathbf{x}_0$.

Now, the closed-loop system model given in the form (4.14) has the terms $\Delta f(\Delta\mathbf{x}, \hat{\mathbf{x}}_0, d)$ and $\Delta G(\Delta\mathbf{x}, d)$, defined by the following expressions:

$$
\Delta f(\Delta\mathbf{x}, \hat{\mathbf{x}}_0, d) = f_1 + f_2^{(1)} + f_2^{(2)} + f_3 + f_4^{(1)} + f_4^{(2)} + f_5^{(1)} + f_5^{(2)} + \psi_r
$$
$$
\Delta G(\Delta\mathbf{x}, d) = \left[\begin{array}{c} 0^{(n \times n)} \\ H^{-1}(I - J^{-T}\hat{J}^T) \end{array} \right];
$$
$$
f_1 = A_1\Delta\mathbf{x} =
$$

$$= \begin{bmatrix} 0^{(n \times n_1)} & 0^{(n \times n_2)} & I^{(n \times n_1)} & I^{(n \times n_2)} \\ Q_1 K_P^{(1)} + Q_1^F \hat{M}_{21} & Q_1^F \hat{M}_{22} & Q_1 K_V^{(1)} - Q_1^F \hat{M}_{21} & -Q_1^F \hat{M}_{22} \end{bmatrix} \Delta \mathbf{x};$$

$$f_2^{(1)} = A_2^{(1)} \dot{\mathbf{x}}_0 =$$

$$= \begin{bmatrix} 0^{(n \times n)} & 0^{(n \times n_1)} & 0^{(n \times n_2)} \\ 0^{(n \times n)} & Q_1 - A_2^{11} & 0^{(n \times n_2)} \end{bmatrix} \dot{\mathbf{x}}_0;$$

$$f_2^{(2)} = A_2^{(2)} \dot{\mathbf{x}}_0 =$$

$$= \begin{bmatrix} 0^{(n \times n)} & 0^{(n \times n_1)} & 0^{(n \times n_2)} \\ 0^{(n \times n)} & A_2^{11} - A_2^{10} & 0^{(n \times n_2)} \end{bmatrix} \dot{\mathbf{x}}_0;$$

$$f_3 = \begin{bmatrix} 0^{(n \times 1)} \\ H^{-1}(J^{-T} \hat{J}^T \hat{h} - h) \end{bmatrix};$$

$$f_4^{(1)} = \begin{bmatrix} 0^{(m \times 1)} \\ -Q_1^1 \hat{L}^{(2)} + Q_5 \hat{M}_{22}^{-1} \hat{L}^{(2)} \end{bmatrix};$$

$$f_4^{(2)} = \begin{bmatrix} 0^{(m \times 1)} \\ -Q_5 \hat{M}_{22}^{-1} \hat{L}^{(2)} + Q_4 \hat{M}_{22}^{0^{-1}} \hat{L}_0^{(2)} \end{bmatrix};$$

$$f_5^{(1)} = A_5^{(1)} F_0^{(2)} =$$

$$= \begin{bmatrix} 0^{(m \times m_1)} \\ Q_5 \hat{M}_{22}^{-1} - Q_1^1 \end{bmatrix} F_0^{(2)};$$

$$f_5^{(2)} = \begin{bmatrix} 0^{(m \times m_1)} \\ Q_4 \hat{M}_{22}^{0^{-1}} - Q_4 \hat{M}_{22}^{-1} \end{bmatrix} F_0^{(2)} = A_5^{(2)} F_0^{(2)};$$

$$\psi_r = \begin{bmatrix} 0^{(n \times 1)} \\ \psi_r^1 \end{bmatrix};$$

$$\psi_r^1 = Q_1^F [-\int_0^t \dot{\hat{M}}_{21} \Delta q^{(1)} dt - \int_0^t \dot{\hat{M}}_{22} \Delta q^{(2)} dt +$$

$$+ \int_0^t (\hat{M}_{21}^0 - \hat{M}_{21}) \ddot{q}_0^{(1)} dt + \int_0^t (\hat{M}_{22}^0 - \hat{M}_{22}) \ddot{q}_0^{(2)} dt +$$

$$+ \int_0^t (\hat{L}_0^{(2)} - \hat{L}^{(2)}) dt];$$

$$n = m = 6; \quad n_1 = n_2 = n/2; \quad m_1 = m_2 = m/2;$$

$$\hat{H} = \begin{bmatrix} \hat{H}_1^{(n \times n_1)} & \hat{H}_2^{(n \times n_2)} \end{bmatrix};$$

$$Q_1 = H^{-1} J^{-T} \hat{J}^T (\hat{H}_1 - \hat{H}_2 \hat{M}_{22}^{-1} \hat{M}_{21});$$

$$Q_1^1 = H^{-1} J^{-T} \hat{J}^T \hat{H}_2 \hat{M}_{22}^{-1};$$

$$Q_1^F = -Q_1^1 K_F^{(2)};$$

$$Q_5 = \begin{bmatrix} 0^{(m_1 \times m_2)} \\ I^{(m_2 \times m_2)} \end{bmatrix}; \quad Q_4 = Q_1^1 \hat{M}_{22}^{0^{-1}};$$

$$A_2^{10} = \begin{bmatrix} I^{(n_1 \times n_1)} \\ -\hat{M}_{22}^{0^{-1}} \hat{M}_{21}^0 \end{bmatrix}; \quad A_2^{11} = \begin{bmatrix} I^{(n_1 \times n_1)} \\ -\hat{M}_{22}^{-1} \hat{M}_{21} \end{bmatrix}; \tag{4.35}$$

The matrices A_1, $A_2^{(1)}$, $A_2^{(2)}$ are of dimensions $(2n \times 2n)$, $A_5^{(1)}$ and $A_5^{(2)}$ of dimensions $(2m \times m_2)$, while ψ_r is a $(2n \times 1)$ vector. The matrices $\hat{H}_1$ and $\hat{H}_2$ represent the corresponding submatrices of the inertia matrix $\hat{H}$, whose elements are defined in [11], whereas the dimensions of these matrices are given by (4.35). The notation I represents the unit matrix of the corresponding rank, given also in (4.35).

A simple interpretation of the terms $f_i^{(\cdot)}$, $i = 1, \ldots, 5$ would be as follows. The expression f_1 contains the terms in the feedback loops (position, velocity, and integral force feedback). This expression depends on the inaccuracy of the model calculation and the values of control gains. The term $f_2^{(1)}$ appears as a consequence of the calculation inaccuracy of the vehicle's model as well as of variations of nominal velocities and accelerations. In the case when $d \equiv d_0$ or $\dot{x}_0 \equiv \underline{0}$, then $f_2^{(1)} \equiv \underline{0}$. The term $f_2^{(2)}$ appears as a consequence of the difference in the calculation of the environment model for the real and nominal values of state coordinates, as well as due to variations of the values of nominal velocity and acceleration. The term f_3 appears because of the inaccuracy of calculation of the centrifugal and gravitational forces of the vehicle model. If $d \equiv d_0$, this term vanishes, i.e. $f_3 \equiv \underline{0}$. The term $f_4^{(1)}$ is a result of the inaccuracies of the overall vehicle-environment model. In the case when $d \equiv d_0$, then $f_4^{(1)} \equiv \underline{0}$. The term $f_4^{(2)}$ is due to the difference of the model calculation on the basis of the real and nominal states of the system. The term $f_5^{(1)}$ is also a result of the inaccuracies in the models of the vehicle and environment, as well as of variations of the nominal values of the function $F_0^{(2)}$. If $d \equiv d_0$ or $F_0^{(2)} \equiv \underline{0}$, then $f_5^{(1)} \equiv \underline{0}$. The term $f_5^{(2)}$ appears as a consequence of the difference in calculation of the model using the real and nominal state quantities. The term $\Delta G \cdot F$ describes the influence of the external forces acting upon the system, and it depends on the inaccuracies of calculating the matrix $\hat{J}$. If $d \equiv d_0$, then $J \equiv \hat{J}$ and $\Delta G \equiv \underline{0}$. The term ψ_r depends on both the value of expressions in the integral force feedback and model inaccuracies.

4.4.2.1 Derivation of Terms of Stability Conditions

If the function $v(t, \Delta x)$ is chosen in the form (4.17), then the derivative of the function v along the solution of the equation (4.14) can be written in the form:

$$
\begin{aligned}
\dot{v} = {} & \frac{1}{2\,v}[\Delta x^T(HA_1 + A_1^T H)\Delta x] + \\
& + (grad(v))^T[A_2^{(1)}\dot{x}_0 + A_2^{(2)}\dot{x}_0 + f_3 + f_4^{(1)} + f_4^{(2)} + \\
& + A_5^{(1)}F_0^{(2)} + A_5^{(2)}F_0^{(2)} + \psi_r] + (grad(v))^T\Delta G(\Delta x)F
\end{aligned}
\tag{4.36}
$$

Estimation of the terms, i.e. the right-hand side functions of (4.36), can be carried out.

Taking into account the aforementioned, the following estimations are obtained:

$$(\text{I}) \qquad \frac{1}{2\,\mathrm{v}}[\Delta\mathrm{x}^T(\mathrm{H}A_1 + A_1^T\mathrm{H})\Delta\mathrm{x}] < -\min_{i,d\in\mathbf{D}}|\sigma_i(A_1)| \cdot \lambda_m^{1/2}(\mathrm{H}) \cdot \|\Delta\mathrm{x}\|$$

$$(\text{II}) \qquad (grad(\mathrm{v}))^T A_2^{(1)}\dot{\mathrm{x}}_0 < \sum_{j=1}^{2n} \xi_j^{(1)}|\Delta\mathrm{x}_j(t)| + \sum_{j=1}^{2n} \xi_j^{(2)}|\dot{\mathrm{x}}_{0_j}(t)| + \xi_0^{(1,2)}$$

$$(\text{III}) \qquad (grad(\mathrm{v}))^T A_2^{(2)}\dot{\mathrm{x}}_0 < \sum_{j=1}^{2n} \xi_j^{(3)}\|\dot{\mathrm{x}}_0(t)\| \cdot |\Delta\mathrm{x}_j(t)|$$

$$(\text{IV}) \qquad (grad(\mathrm{v}))^T f_3 < \xi^{(4)} + \xi^{(5)}\|\Delta\mathrm{x}(t)\|$$

$$(\text{V}) \qquad (grad(\mathrm{v}))^T f_4^{(1)} < \xi^{(6)} + \sum_{j=1}^{2n} \xi_j^{(7)}|\Delta\mathrm{x}_j(t)|$$

$$(\text{VI}) \qquad (grad(\mathrm{v}))^T f_4^{(2)} < \sum_{j=1}^{2n} \xi_j^{(8)}|\Delta\mathrm{x}_j(t)|$$

$$(\text{VII}) \qquad (grad(\mathrm{v}))^T A_5^{(1)} F_0^{(2)} < \sum_{j=1}^{2n_2} \xi_j^{(9)}|\Delta\mathrm{x}_j(t)| +$$

$$+ \sum_{j=1}^{m_2} \xi_j^{(10)}|F_{0_j}^{(2)}(t)| + \xi_0^{(9,10)}$$

$$(\text{VIII}) \qquad (grad(\mathrm{v}))^T A_5^{(2)} F_0^{(2)} < \sum_{j=1}^{2n} \xi_j^{(11)}\|F_0^{(2)}(t)\| \cdot |\Delta\mathrm{x}_j(t)|$$

$$(\text{IX}) \qquad (grad(\mathrm{v}))^T \psi_r(t) < \sum_{j=1}^{2n} \xi_j^{(12)}\|\dot{\mathrm{x}}_0(t)\| \cdot |\Delta\mathrm{x}_j(t)| +$$

$$+ \sum_{j=1}^{2n} \xi_j^{(13)}|\Delta\mathrm{x}_j(t)| + \xi_0^{(13)}$$

$$(\text{X}) \qquad (grad(\mathrm{v}))^T \Delta G(\Delta\mathrm{x}) F < \sum_{j=1}^{m} \xi_j^{(14)}|F_j(t)| +$$

$$+ \sum_{j=1}^{2n} \xi_j^{(15)}|\Delta\mathrm{x}_j(t)| + \xi_0^{(14,15)} \qquad (4.37)$$

where σ_i, $i = 1, 2, \ldots, 2n$, are the eigenvalues of the matrix A_1 for the given parameters d; λ_m is the minimal eigenvalue of the positive definite matrix H; γ is a scalar value calculated from the expression $\gamma = min|\sigma_i| \cdot \lambda_m^{1/2}(\mathrm{H})$; $\xi_j^{(k)}$ are positive or negative coefficients in the estimation functions for $k = 1, \cdots, 15$; x_j denotes the j-th component of the vector x; the inequalities (4.37) should be valid for $\forall \mathrm{x} \in \mathbf{X}^t(t)$, $\forall t \in \mathbf{T}$ and $d \in \mathbf{D}$.

By substituting the expressions from (4.37) into the expression for the function $\dot{\mathrm{v}}$ derivative, and having in mind the condition (4.6), a concrete form of the function $\Psi(t)$ is obtained as:

$$
\begin{aligned}
\Psi(t) \;=\; & -\gamma\, \underline{X}^I exp(-\alpha t) + \\
& + \; (\xi^{(1)} + \xi^{(5)} + \xi^{(7)} + \xi^{(8)} + \xi^{(9)} + \xi^{(13)} + \xi^{(15)})\underline{X}^t exp(-\alpha t) + \\
& + \; (\xi^{(4)} + \xi^{(6)} + \xi_0^{(1,2)} + \xi_0^{(9,10)} + \xi_0^{(13)} + \xi_0^{(14,15)}) + \\
& + \; (\xi^{(3)} + \xi^{(12)})\|\dot{x}_0(t)\|\underline{X}^t exp(-\alpha t) + \\
& + \; \sum_{j=1}^{2n} \xi_j^{(2)} |\dot{x}_{0_j}(t)| + \xi^{(11)}\|F_0^{(2)}(t)\|\underline{X}^t exp(-\alpha t) + \\
& + \; \sum_{j=1}^{m_2} \xi_j^{(10)} |F_{0_j}^{(2)}(t)| + \sum_{j=1}^{m} \xi_j^{(14)} |F_{0_j}(t) + \underline{F}_{-j}^t exp(-\beta t)| \qquad (4.38)
\end{aligned}
$$

where $\xi^{(1)}$, $\xi^{(5)}$, $\xi^{(7)}$, $\xi^{(8)}$, $\xi^{(9)}$, $\xi^{(13)}$, and $\xi^{(15)}$ are maximal values of the corresponding components $\xi_j^{(k)}$, $k = 1, 5, 7, 8, 9, 13, 15$. $\underline{F}_{-j}^t$ are positive numbers corresponding to the region $\mathbf{F}^t(t)$ for the index values $j = 1, \ldots, m$. Taking into account the relation (4.38), the general condition for practical stability (4.7) can be finally written in the form:

$$
\begin{aligned}
\int_0^t \Psi(t')dt' \;=\; & [-\hat{\gamma}\, \underline{X}^I + (\xi^{(1)} + \xi^{(5)} + \xi^{(7)} + \xi^{(8)} + \\
& + \; \xi^{(9)} + \xi^{(13)} + \xi^{(15)})]\underline{X}^t \frac{(1 - exp(-\alpha t))}{\alpha} + \\
& + \; [(\xi^{(3)} + \xi^{(12)})\,\dot{x}_0^u + \xi^{(11)} F_0^u]\underline{X}^t \frac{(1 - exp(-\alpha t))}{\alpha} + \\
& + \; (\xi^{(4)} + \xi^{(6)} + \xi_0^{(1,2)} + \xi_0^{(9,10)} + \xi_0^{(13)} + \xi_0^{(14,15)})\, t + \\
& + \; \sum_{j=1}^{2n} \xi_j^{(2)} \int_0^t |\dot{x}_{0_j}(t)|dt + \sum_{j=1}^{m_2} \xi_j^{(10)} \int_0^t |F_{0_j}^{(2)}(t)|dt + \\
& + \; \sum_{j=1}^{m} \xi_j^{(14)} (\int_0^t |F_{0_j}(t)dt| + \underline{F}_{-j}^t \frac{(1 - exp(-\beta t))}{\beta}) < \\
& < \; \lambda_m^{1/2}(\mathbf{H})\underline{X}^t\, exp(-\alpha t) - \lambda_M^{1/2}(\mathbf{H})\underline{X}^I \qquad (4.39)
\end{aligned}
$$

The following notations are used: $\dot{x}_0^u$ denotes a positive number such $\|\dot{x}_0(t)\| < \dot{x}_0^u$ and F_0^u is also a positive number such that $\|F_0(t)\| < F_0^u$; $\hat{\gamma}$ denotes a number corresponding to the number γ but which also includes all the negative numbers amongst $\xi_j^{(1)}$, $\xi_j^{(5)}$, $\xi_j^{(7)}$, $\xi_j^{(8)}$, $\xi_j^{(9)}$, $\xi_j^{(13)}$, $\xi_j^{(15)}$ and also amongst $\xi_j^{(3)} \cdot \dot{x}_0^u$, $\xi_j^{(12)} \cdot \dot{x}_0^u$, $\xi_j^{(11)} \cdot F_0^u$. Taking into account the expressions (4.36), (4.38) and (4.39), the expressions for the stability conditions (4.6) and (4.7) can be written for the case of the combined position-force control law (3.8).

4.5 Implementation Aspects

In order the system be stable it is necessary to synthesize the control gains $K_P^{(1)}$, $K_V^{(1)}$, and $K_F^{(2)}$ in such a way that they are able to compensate for the effects of the inaccuracies in modeling, as well as for the effect of the uncertainties of parameters during the vehicle motion. It is necessary to schedule such gains in the controller that will in a best way adapt the control system to the uncertainties existing in the system. The problem is solved in such a way that the sets of system control gains are calculated off-line for various possible motion conditions, as well as for the different expected values of the system parameters and tire-road interaction parameters. For various motion conditions, stability conditions are checked using a procedure that will be presented in Section 4.6. To this purpose the values of control gain matrices used in the control law (3.8) should be varied in order the system satisfies the set of stability conditions (4.6) and (4.7). In this way one can determine the extent to which the system's inaccuracies influence the vehicle's stability. Based on the stability conditions derived in this chapter, an automatic procedure can be developed which will serve in the process of control design to detect the satisfactory values of the control gains for which the system fulfils the stability conditions. The gains thus determined are then scheduled in the memory of the vehicle controller in a tabular form and used afterwards in the real-time operation.

The dynamic controller proposed in the papers [12, 13], and further expanded in Chapter 3 of this monograph, can be designed to operate as a multiprocessor-multitask system that can perform simultaneously several operating functions in parallel regime. These are: (i) real-time estimation of the model parameters and tire-load interaction parameters, (ii) calculation of control signals on the higher, tactical control level, and (iii) calculation of the local control signals at the executive level of the vehicle's actuators.

As for the practical values of the synthesized vehicle controller, it should be pointed out that the described dynamic controller can guarantee a stable behavior of the system and its robustness against the parametric uncertainties in their finite range of limits. Perturbations of larger magnitudes, as well as variations of parameters in a wider range, cannot be compensated for in a satisfactory way using this kind of control. Besides, it has been assumed that the structure of the models of the system and its environment used here can approximate well the real vehicle-environment complex system. Otherwise, the proposed controller has to be upgraded by adding a supplementary neuro compensator that would compensate for the effects of the unmodeled system dynamics and those perturbations which were not encompassed by the model. The choice of such a neuro-compensator, based on the topology of a multilayered NN would resolve the possible technical constraints of the proposed dynamic controller. This was also described in Chapter 3 of this monograph.

Now we will describe in brief the way of scheduling control gains for practical implementation into the proposed dynamic controller of the road vehicle. The main goal in estimating the values of control gains is to ensure a stable behavior of the vehicle, i.e. the adaptive action of the controller to variable parameters of the vehicle and its environment. How is this practically achieved? Gains of the synthesized controller are calculated using the developed test for checking the vehicle stability conditions in dependence of the dynamic law applied in control. As described in the previous section, the controller gains are determined on the basis of checking stability conditions for the assumed set of parameters $\hat{d}$ of the system and its environment and for an arbitrarily selected object trajectory $x_0(t)$. Values of the assumed parameters are most often different from the real values of the system parameters d. For this reason the calculated gains are valid only for one nominal trajectory and for the assumed values of model parameters. For another trajectory and another set of values of the parameters of the system and environment, it is not possible to guarantee practical stability of the system until stability conditions are tested again. It should be pointed out that relatively small variations of the model parameters (in the range of $2 - 3$ %), as well as relatively small geometrical deviations of the given nominal system trajectories, do not essentially affect behavior of the vehicle during its motion. One can take advantage of this fact in the following way. Using a computer, elements of the matrices of control gains are determined automatically for the different combinations of the possible values of parameters of the system model and for the different types of vehicle trajectories. Hereby, to test the algorithm use is mainly made of three characteristic paths of vehicle motion: straight line, curvilinear path (of different curvature radius), and sinusoidal path (of various amplitudes and periods). These paths, as well as their combinations, can approximate the majority of road vehicle trajectories in real exploitation. The gain matrices obtained by checking the conditions of practical system's stability are stored in the controller memory at the given addresses as constant tabulated values. At the same time, the corresponding values of the system parameters $\hat{d}$ and geometrical pathway parameters for which the control gains were determined are also stored. Theoretically, the off-line procedure of determining controller gains could be repeated unlimited number of times, because there is a large number of possible combinations of values of the parameters which are taken into consideration. In practice, however, in the algorithm implementation it suffices to repeat the procedure several hundreds of times using certain selected combinations of parameters from the possible set of real values of D. The mode of real-time operating of the dynamic controller was described in Section 3.5 for one, potentially possible, controller architecture. On the basis of the estimated values of unknown model parameters[4], the synthesized controller is capable of "recognizing" a most similar combination of the parameter values of the system model. For that variant, the controller is to "load" from the memory the corresponding gain matrices for the current vehicle path, which have been already prepared and saved in the form of tabulated values. In this way, a certain adaptability of the dynamic control law to the variable motion conditions is achieved. The proposed variant of architecture of the dynamic controller is

---

[4]Note that estimation of the parameters of the system and its environment should be carried out in real time.

shown in Fig. 3.21.

At the end of this section let us add an important notice that the so-called scheduling of the control gains K_P, K_V and $K_F^{(2)}$ by this procedure cannot guarantee an absolutely stable system behavior for all the possible system trajectories and for all possible values of the parameters from the set D. This is a logical consequence since, for the purely technical reasons, it is not possible to check in reality the conditions of practical stability for the absolutely all possible combinations of parameters of the system and its environment around the nominal trajectory (Fig. 4.2). For this reason, in the preceding chapter we proposed the introduction of an additional neuro-compensator to the control system. Its integration into the so-called hybrid neuro-dynamic vehicle controller should ensure a higher robustness of the automated control system with respect to the existing model inaccuracies. This additionally improves the system's controllability and contributes to an enhanced motion stability.

4.6   Simulation Example

An essential task of the control system designer is to answer the following pragmatic question. Can the vehicle practical stability be guaranteed even in the presence of system's uncertainties and when it is possible to predict the real range of variation of the parameters dominantly affecting the system performance? In the present example we will show that an appropriate selection of control gains can yield the indispensable system stability, as well as the robustness of the control system with respect to the inevitable inaccuracies of the model applied.

In this simulation example we consider the vehicle stability along an arbitrary curvilinear trajectory $x_0(t)$, $t \in (0,8]$ $[sec]$. The real parameters (geometrical and dynamic) of the vehicle used to evaluate stability conditions and model simulation were taken from Appendix of Chapter 2. As far as the vehicle parameters used in the models (2.28) and (2.73) are concerned, the most influential of them upon the system's behavior are: vehicle mass, moments of inertia, relative position of the roll and pitch axes with respect to the vehicle MC in particular directions of motion, rolling and sliding coefficients of tire pneumatics, aerodynamic resistance in the longitudinal and lateral directions of motion, yaw damping coefficient about the vehicle vertical axis, cornering stiffness, etc. These parameters have a dominant influence on the dynamic system performance. For this reason, their influence upon the vehicle stability has to be analyzed. In reality, in the course of vehicle motion, some of these parameters can change their values to a significant extent. This refers before all to the mass, moment of inertia, and friction coefficients, whereas the geometrical and aerodynamic resistance coefficient can be regarded as invariable quantities, determined by the vehicle design. Their variation can essentially deteriorate the quality of the system's behavior and, finally, yield an unstable behavior. Bearing in mind the aforementioned, a real case (but not an extreme one) was considered in the current example involving the changes of the vehicle body mass and moment of inertia during the motion in a range of $\pm 10\%$ w.r.t. their expected values. It was taken into account that the change of the vehicle mass is accompanied by a change in the relative position of the structure MC, as well as of the positions of the roll and pitch centers of the vehicle body about the corresponding axes. Also, the influence of a decrease in tire sliding coefficient is considered, assuming they were lowered in the range $0-30\%$ compared with the vehicle motion on a clean dry road surface. In the current example we selected the vector of vehicle model parameters d of dimensions (18×1), which takes its values from the set D of the possible parameter values. It can be represented by the following expression:

$$d = \begin{bmatrix} m & I_x & I_y & I_z & h_1 & h_2 & h_3 & h_4 & h_5 & f_r & K_z \cdots \\ \nu_{s_1} & \nu_{s_2} & \nu_{s_3} & \nu_{s_4} & \nu_{d_1} & \nu_{d_2} & \nu_{d_3} & \nu_{d_4} \end{bmatrix}^T \tag{4.40}$$

The adopted values and meanings of the symbols are given in Appendix of Chapter 2. In the examining stability conditions it was supposed that the elements of the vector d have variable values during motion.

In the current simulation example, initial deviation of the state vector $\Delta x(0)$ from the nominal vales were arbitrarily selected, so that they have relatively small finite values represented by the vector:

$$\begin{aligned} \Delta(x)(0) \quad = \quad & [-0.200 \quad -0.100 \quad -0.030 \quad +0.005 \ \cdots \\ & \cdots \quad +0.010 \quad -0.010 \quad -0.050 \quad -0.010 \cdots \\ & \cdots \quad -0.005 \quad +0.0005 \quad +0.001 \quad -0.001]^T \end{aligned} \tag{4.41}$$

It is known that the nominal state vector (4.4) is of dimensions (12×1) and hence the vector $\Delta(x)(0)$ represents the initial deviations of the vehicle body position and velocity in particular motion directions. Also, it was adopted that the value of the nominal state vector $x_0(0)$ in the initial time moment $t = 0$, is:

$$x_0(0) = [0 \ \ 0 \ \ 0 \ \ +0.6 \ \ 0 \ \ 0 \ \ 22.224 \ \ 0 \ \ 0 \ \ 0 \ \ 0 \ \ 0]^T. \tag{4.42}$$

As can be seen, the initial nominal state coordinates are zero, except for the relative position of the vehicle body MC with respect to the road surface, which has a finite value $z_0(0) = 0.6$, and nominal speed of vehicle motion in the longitudinal direction $\dot{x}_0(0) = 22.224 \ [m/s]$. This motion corresponds to a constant linear vehicle speed of $V_0 = 80 \ [km/h]$ realized in a stationary regime of motion. The real value of the state vector in the initial moment $x(0)$ is determined from the expression:

$$x(0) = x_0(0) + \Delta x(0) \tag{4.43}$$

which represents the so-called initial (i.e. perturbed) system state. Dimensions of the finite region X^I, defined in Fig. 4.2, were given so that the chosen initial state $x(0)$ belongs to the mentioned region $x(0) \in X^I$ (see Fig. 4.2) and that the Euclidean norm of the state vector in the initial time instant satisfies the inequality $\|\Delta x(0)\| < \underline{X}^I$. In the considered simulation example, an arbitrary size of the region X^I was adopted (see Fig. 4.2) so that $\mathbf{X^I} = 1.01 \cdot |\Delta x(0)|$. In agreement with the control task definition given in Section 4.3, such vector $\mathbf{X}^t(0)$ was selected whose elements by their absolute value exceed the corresponding elements given in the vector $\mathbf{X}^I$. Concretely, the selected values in the given vector are:

$$\begin{aligned}
\mathbf{X}^t(0) \quad = \quad & [0.22 \quad 0.12 \quad 0.051 \quad +0.007 \quad 0.045 \quad 0.034 \cdots \\
& 0.1 \quad 0.1 \quad 0.05 \quad 0.003 \quad 0.01 \quad 0.01]^T
\end{aligned} \tag{4.44}$$

Each of the elements $(j = 1, \ldots, 12)$ of the vector $\Delta\mathbf{x}_j(0)$ has the property that $|\Delta\mathbf{x}_j(0)| < \mathbf{X}^I_j < \mathbf{X}^t_j(0)$.

In order the considered system would possess a desired stability rate (rate of motion convergence) the value of stability exponent α was assigned. As is known, deviations of the state quantities $\Delta\mathbf{x}(t)$ over the considered time interval $t \in (0, 8]$ $[s]$ should be within the prescribed stability region $\|\Delta\mathbf{x}(t)\| < \underline{X}^t exp(-\alpha t)$. Because of that, to this exponent was assigned a positive constant value $\alpha = 1.7269$, determined so as the deviation norm of the state vector $\|\Delta\mathbf{x}(t)\|$ after $t = 8$ $[s]$ of motion would drop below the desired error limit.

In the current example, the combined position-force control algorithm (3.8) was applied. For this purpose, the model parameters (2.73) of the "dynamic environment" were estimated by using conventional 3D-model[5] (see Chapter 2). A first step in deriving the concrete expressions for the stability conditions (4.6) and (4.7) was the determination of coefficients in the estimation functions (4.37). First of all, it was necessary to determine the state matrix A_1 and the positive definite matrix H in such a way that the matrix inequality (I) from (4.37) is satisfied. This requirement reduces to a simultaneous calculation of the matrices A_1 and H, applying a software procedure specially developed for this purpose. The procedure consists essentially of iterative calculation of the corresponding matrix inequality for the imposed, randomly generated gain values, whereby the conditionally "whole state space" along the system trajectory $\mathbf{x}_0(t)$ (Fig. 4.2) is taken into account. Since all the possible values from the observed "tubular" state space (Fig. 4.2) cannot be taken into account only a finite number of samples is considered. The procedure is terminated when the inequality (I) from (4.37) is satisfied, whereby the real parts of the eigenvalues of the matrix A_1 are negative and that the matrix H is a positive definite matrix. For the matrices A_1 and H thus determined, the control gains $K_P^{(1)}$, $K_V^{(1)}$ and $K_F^{(2)}$ in the control law (3.8) are calculated as:

$$K_P^{(1)} = \begin{bmatrix} 56.84 & 3.77 & 6.07 \\ 8.54 & 127.91 & 10.14 \\ 17.32 & 3.09 & 88.82 \end{bmatrix} ; \quad K_V^{(1)} = \begin{bmatrix} 12.06 & 0.81 & 1.29 \\ 1.81 & 18.09 & 2.15 \\ 3.67 & 0.65 & 15.08 \end{bmatrix} ;$$

$$K_F^{(2)} = \begin{bmatrix} 2.99 & 0.01 & 0.01 \\ 0.10 & 2.99 & 0.04 \\ 0.07 & 0.02 & 2.99 \end{bmatrix} \tag{4.45}$$

---

[5]Another possibility is to use the measurement results obtained on the experimental platform shown in Fig. 2.26.

For these gains, the eigenvalues of the matrix A_1 have all negative real parts, while the positive definite matrix H has all positive eigenvalues. They are illustrated in Fig. 4.3a.

If the control gains $K_P^{(1)}$, $K_V^{(1)}$ and $K_F^{(2)}$ are, for example, selected arbitrarily by the system designer in the way described by the following term:

$$K_P^{(1)} = \begin{bmatrix} 1 & 1 & 1 \\ 1 & 1 & 1 \\ 1 & 1 & 1 \end{bmatrix} ; \quad K_V^{(1)} = \begin{bmatrix} 0.1 & 0.1 & 0.1 \\ 0.1 & 0.1 & 0.1 \\ 0.1 & 0.1 & 0.1 \end{bmatrix} ;$$

$$K_F^{(2)} = \begin{bmatrix} 0.3 & 0.3 & 0.3 \\ 0.3 & 0.3 & 0.3 \\ 0.3 & 0.3 & 0.3 \end{bmatrix} \tag{4.46}$$

then the eigenvalues of the state matrix A_1 will involve some positive members too, as is shown in Fig. 4.3b. In that case, the system will exhibit unstable behavior, which will be verified by the corresponding simulation results. For the control gains defined by (4.45), the system considered has stable characteristics and its motion in the state space converges to the nominal trajectory within the prescribed stability regions (see Fig. 4.4). On the other hand, in the case when the control gains (3.8) are selected in the arbitrary way defined by (4.46), the system exhibits unstable behavior and its motion diverges in the state space, which is illustrated in Fig. 4.5.

By analyzing the obtained simulation results presented in Figs. 4.4 and 4.5, it can be concluded that the control algorithm (3.8) ensures stable behavior of the vehicle in the six main coordinate direction of motion in every time instant over the interval $t \in (0, 8]$ $[s]$ when the control gains are defined as in (4.45). Nevertheless, though the simulation results from Fig. 4.4 indicate a stable behavior of the system for the supposed parameter variation, the stability conditions (4.6) and (4.7) have to be properly checked out, to prove the system's stability for all the possible values of the parameters varying in the expected range. In order to do it, the first step is to determine the estimation function in (4.37). The simplest way to identify these estimations is to define them in a linear form, but the estimates of the functions can also be of a nonlinear character if they better approximate the function than their linear competitors. An example of a linear and potential nonlinear estimation function of $((grad(\mathrm{v}))^T \cdot f_3)$ from (4.37) is illustrated in Fig. 4.6. As can be seen from Fig. 4.6, the coefficients $\xi^{(4)}$ and $\xi^{(5)}$ can be the ones defining the analytical form of a straight line or a certain parabolic function, for example. Here we assume only a linear form of the estimation terms (4.37). The unknown coefficients $\xi_j^{(i)}$ in the estimation functions (4.37) can be determined in a relatively simple manner by using an iterative software "searching" procedure. In fact, by varying argument of any of the functions on the left-hand side of each term existing in the set of expressions (4.37), the function increments on the left-hand side of the inequalities are calculated. This is done for finite, small increments of the function's argument in every iterative cycle. The values of the coefficients from the estimation functions are calculated as a quotient of the increment of a function and the absolute

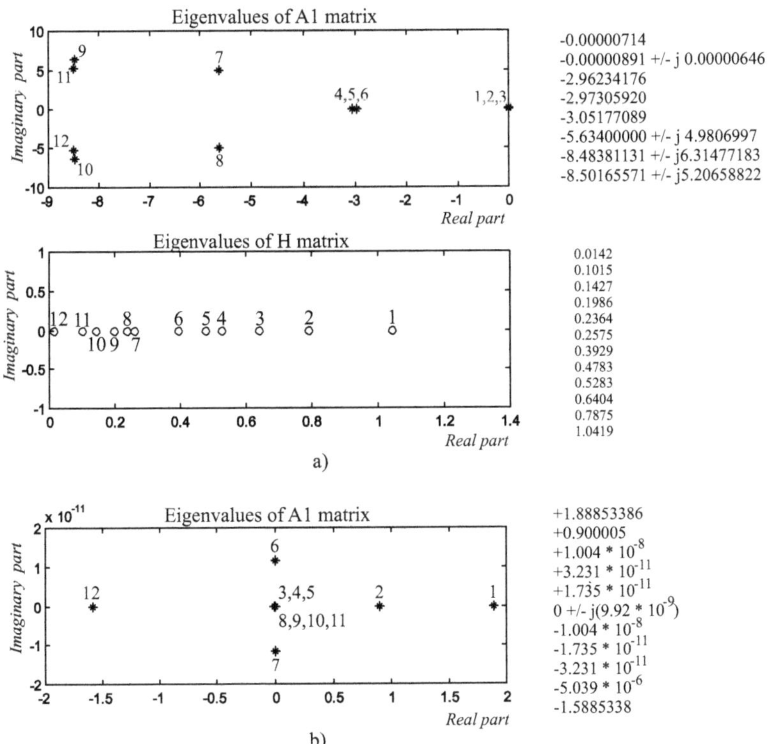

Figure 4.3: a) Eigenvalues positions of the state matrix A_1 in complex plane for control gains defined by expression (4.45) and the corresponding eigenvalues of the positive-definite matrix $\mathbf{H}$, b) Eigenvalues of the state matrix A_1 in the case of control gains selected in the arbitrary manner defined by (4.46)

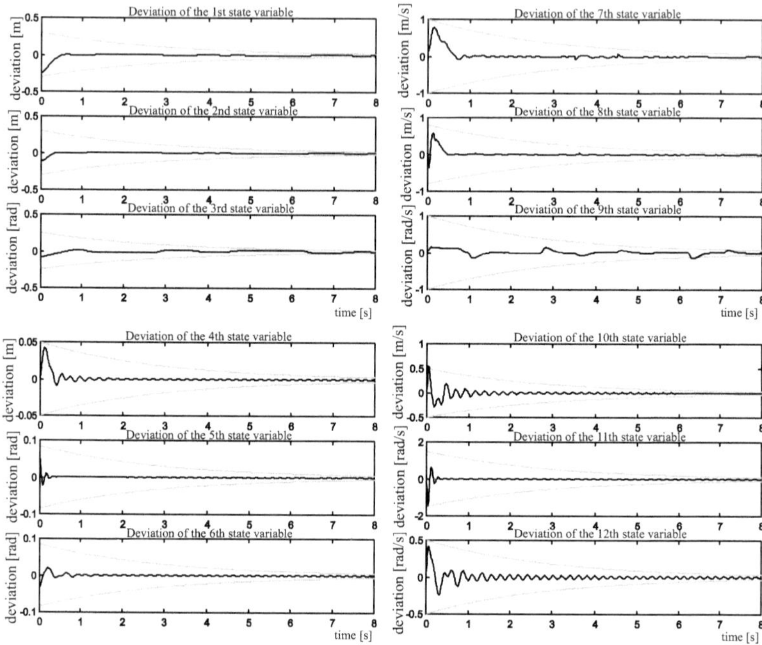

Figure 4.4: Stability indices of the vehicle behavior in the state space w.r.t imposed stability regions for the control gains defined by (4.45)

value of its argument. Thus, a set of points in the space of arguments is obtained that makes the so-called "point picture" (see Fig. 4.6). The definite values of the coefficients in the estimation functions represent "the best solution" that can be determined both analytically and graphically as, for example, the ones illustrated in Fig. 4.7. The straight lines thus obtained represent rather rough approximations of the considered functions. These lines were drawn in such a way that all the points presented in the graphs are "below" w.r.t. their ordinate to the line representing a graphical presentation of the inequality solution. Having in mind the above discussion related to the current example, the results of credibility check of the set of inequalities (II) - (X) from (4.37) are elaborated and presented by the graphs given in Fig. 4.8.

In the case when the gains are appropriately defined and when all the estimation functions (4.37) are successfully identified, then the stability conditions (4.6) and (4.7) can be finally checked out for any time instant over the interval $t \in (0, 8]\ [s]$. The results of the practical stability test for the road vehicle param-

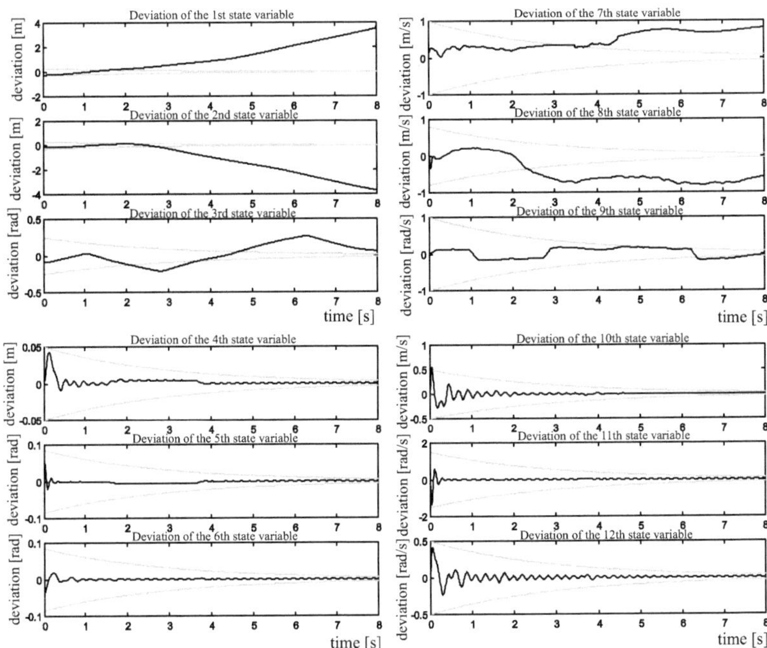

Figure 4.5: Stability indices of the vehicle behavior in the state space w.r.t. imposed stability regions of the control gains defined by (4.46)

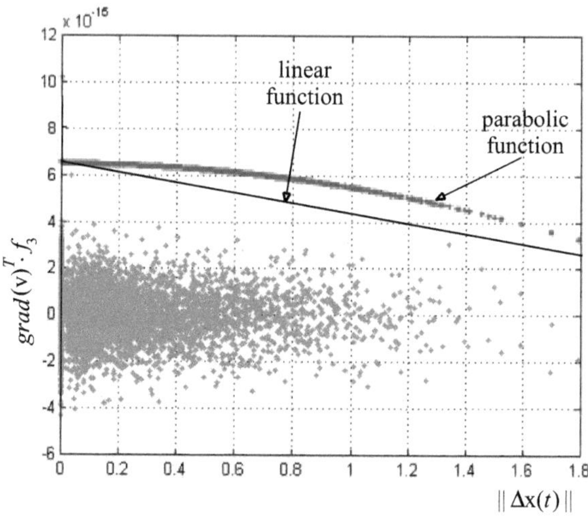

Figure 4.6: Example of a linear and nonlinear form of the estimation function (IV) from the set of expressions (4.37)

eters and conditions defined in the current example are graphically presented in Fig. 4.9.

It can be noticed that the general stability conditions (4.6) and (4.7) are fully satisfied in any time instant $t \in (0,8]$ $[s]$ and in the whole constrained state space along the nominal trajectory (Fig. 4.2).

Bearing in mind the previous discussion concerning the identification of the estimation functions existing in the terms (4.21) and (4.37), and knowing that the determination of coefficients in these functions is the most delicate phase in checking stability conditions, some additional explanations may be needed. In order to make the procedure for evaluating stability test more understandable from the viewpoint of practical implementation, we will give some guidelines that should be taken into account when executing the procedure for identification of estimation functions.

- Estimation of the basis functions[6] is carried out so that the approximation function (estimation function[7]) be dependent on the same variables as the

---

[6]By this notion we understand the corresponding expression on the left-hand side of the inequalities in the relations (4.21) and (4.37), whose approximation (i.e. estimation) we want to determine.

[7]Estimation function is the expression appearing on the right-hand side of the inequalities in the relations (4.21) and (4.37).

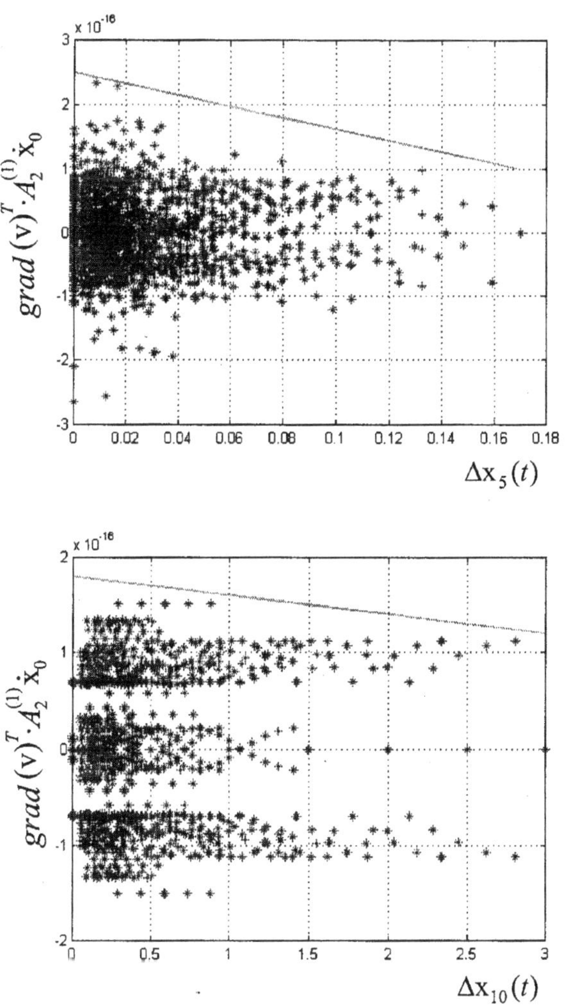

Figure 4.7: Graphical presentation of identification of some of the estimation functions from the set of expressions (4.37)

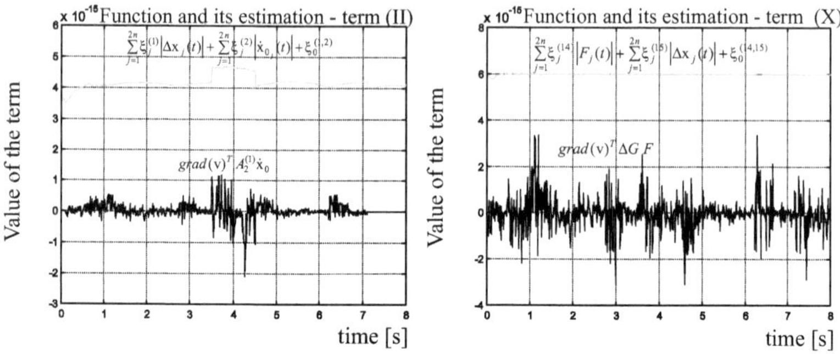

Figure 4.8: Graphical presentation of a function and its estimation (approximation) for two expressions from (4.37)

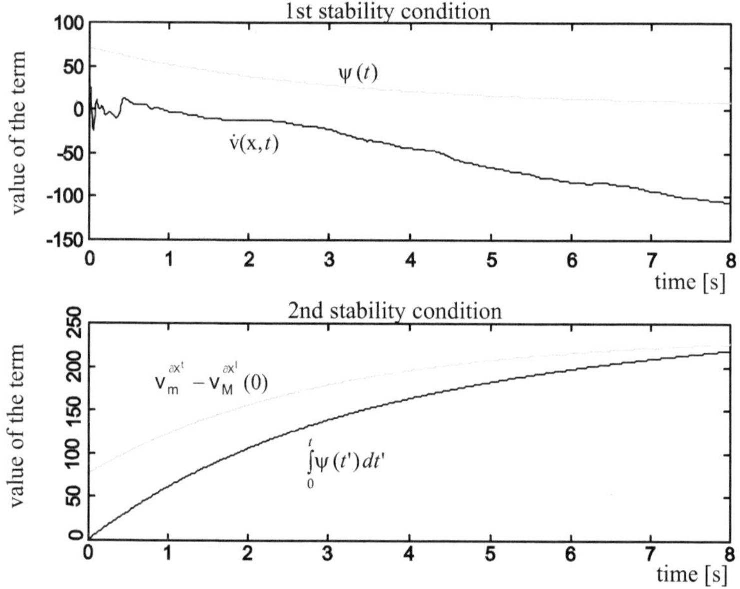

Figure 4.9: Graphical presentation of practical stability test for selected vehicle parameters and conditions considered in the current example

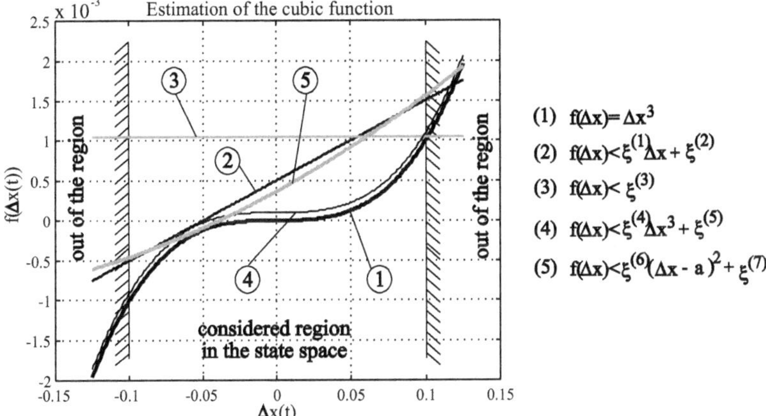

Figure 4.10: Example of estimation of the cubic function $f(\Delta x) = \Delta x^3$ by particular characteristic approximation functions in the given region of change of the argument Δx

basic function. What does it practically mean? If we assume one concrete form of a scalar function denoted by $f(\Delta x)$, then its estimation may be determined in a suitable form, e.g. $f(\Delta x) < \xi^{(1)}\Delta x + \xi^{(2)}$. Obviously, the basis function $f(\Delta x)$ and its estimation $\xi^{(1)}\Delta x + \xi^{(2)}$ are dependent of the same variable Δx. The coefficients $\xi^{(1)}$ and $\xi^{(2)}$ are constants whose values have to be identified.

- By their structure, estimation functions may be linear or nonlinear. They are selected on the basis of experience, and only in a special case when a concrete analytical form of the function is known, the choice of the estimation function is simplified. For example, if we know that the function is of the form $f(\Delta x) = \Delta x^3$, then we can select its estimation in a linear form $\xi^{(1)}\Delta x + \xi^{(2)}$. Hereby, the inequality $f(\Delta x) < \xi^{(1)}\Delta x + \xi^{(2)}$ must be satisfied over the predetermined interval $\Delta x \in [-x^t, +x^t]$. This practical example is illustrated in Fig. 4.10.

- A function can also be approximated by a constant value (see Fig. 4.10). In that case it holds that $f(\Delta x) < \xi^{(3)}$. This is the simplest case of estimation and it represents the most "rough" (most approximate) possible estimation of the basis function.

- By its real absolute value, the estimation of a function must be greater than the absolute value of the basis function in any time instant considered. The coefficients $\xi_j^{(k)}$ should be determined in such a way that the function estimation be as "close" as possible to the function which is estimated. In the example illustrated in Fig. 4.10, this concretely means that the best estimation of the function $f(\Delta x)$ is a certain cubic function which satisfies the condition $f(\Delta x) < \xi^{(4)}\Delta x^3 + \xi^{(5)}$. If we know analytical form of the basis function, then it is convenient to select the estimation function of a similar form. The coefficients in the estimation function should be determined so as the preset condition demanding the estimation function be greater by its absolute value than the basis function in the considered region of possible values of the argument Δx is satisfied.

- In the case of estimation of a vector function (when Δx is say the vector of state coordinate deviation), a characteristic form of estimation function can be defined as $f(\Delta x) < \sum_{j=1}^{n} \xi_j^{(1)}\Delta x_j + \xi^{(2)}$, where $\xi_j^{(k)}$ are the coefficients which can be arbitrary positive or negative numbers. On the basis of experience it can be recommended that the values of these coefficients be negative if possible, for the reason of a higher degree of system stability [14].

- In the search for a most suitable structure of the estimation function, in the stage of checking the predetermined stability conditions, one should not consider a "too large" region $\underline{X}^t$ (Fig. 4.2). It is recommendable that the given region of deviation of the state coordinates x from their nominal values x_0 must not be too large. Usually, certain small values for the boundaries of the state coordinate deviation are given, within which is plausible to expect the system can be stabilized by applying one of the proposed control laws.

- The coefficients $\xi_j^{(k)}$ are determined by an iterative procedure. The procedure will be described in brief on a hypothetical example. Let us assume that a certain function can be estimated in the following way: $f(\Delta x, x_0) < \sum_{j=1}^{n} \xi_j^{(1)}|\Delta x_j(t)| + \sum_{j=1}^{n} \xi_j^{(2)}x_{0j}(t) + \xi^{(3)}$. The coefficients $\xi_j^{(1)}$, $\xi_j^{(2)}$ and $\xi^{(3)}$ are determined sequentially, one by one, by seeking "the best" solution which satisfies the predetermined criterion defined by the inequalities (4.21) and (4.37). Hereby, the "search" is carried out in a tubular region of the state space, with a selected step (increment). Also, in equal time increments data are sampled from the nominal system trajectory over the time interval $t \in (t_0, t_1]$. Values of the vector $\Delta x(t)$ are iteratively varied inside the allowed boundaries $\mathbf{X^t}(t)$ defined by the control law (see Fig. 4.2). The procedure itself is carried out in the following way. In the beginning fulfillment of the expression defined by the corresponding inequality is checked, say for an equilibrium value $\Delta x_1 = 0$. This value represents the first component, i.e. an element of the vector Δx. The procedure is continued by iteratively changing Δx_1 within the boundaries $-x_1^t \leq \Delta x_1 \leq +x_1^t$ with an arbitrary constant increment. The sampling increment must not be too large in order to take into consideration a largest possible number of points from the state space. On the other hand, the increment should not be too small because the numerical process of "searching" (i.e. the software procedure) would take too long time and the procedure would lose its practical applicability. The procedure essentially consists of iterative variation of $\Delta x_1(t)$ so that it is taken for practically all the values of samples from the region $X^t(t)$, along the whole nominal trajectory of the system. This means that the value $\Delta x_1(t = 0)$ is incremented for a finitely small value $\Delta x_1(\Delta t)$ and then $f(\Delta x(t))$ is calculated. At that, the other state quantities $\Delta x_j(\Delta t)$, $j = 2, \ldots, 12$ are "kept" constant. Then $\Delta f = f(\Delta x_1(t + \Delta t)) - f(\Delta x_1(t))$ is calculated. After that, it is necessary to determine the quotient $\Delta f / |\Delta x_1(\Delta t) - \Delta x_1(0)|$. Value of the calculated quotient represents the value of the coefficient $\xi_1^{(1)}(t)$ in the instant t. The procedure is repeated for each time sample. At the end of the procedure only one value $\xi_1^{(1)}(t)$ is selected which from the set of the determined values $\xi_1^{(1)}$ satisfies the given inequality, i.e. the criterion of estimating the function for the selected time instant t and the values of the state coordinate deviation $\Delta x_1(t)$. The procedure is repeated for all other $2n$ components of the vector of state deviation $\Delta x_j(t)$, $j = 1, \ldots, 2n$. As in each cycle is calculated the value of the j-th coefficient $\xi_j^{(1)}$, this value is eventually adopted as the coefficient having the smallest real absolute value (i.e. the most negative value) of all the previously calculated. The fulfillment of the predetermined condition ensures the obtained estimation function be "closest" to the basis function $f(\Delta x)$. As for the determination of the coefficient $\xi_j^{(2)}$, the procedure is the same. The only difference is in that that instead of the "search" of the state space (Fig. 4.2) and the corresponding calculations, the coefficient $\xi_j^{(2)}$ is computed by taking into consideration equidistant time samples from the nominal trajectory $x_0(t)$. When considering the determination of the coefficient $\xi^{(3)}$ in the previous example, we can say the following. If the aim is to bring the system into the zero-th equilibrium state $\Delta x = \underline{0}$ (zero vector), then the value of the vector function is usually $f(\Delta x) = \underline{0}$. If this is not the case, i.e. when $f(\Delta x = 0) \neq \underline{0}$, it is necessary to estimate the "rest" by a function which depends on: (i) time t, or (ii) the trajectory $x_0(t)$, or (iii) variation of the parameter values $\Delta d = d - \hat{d}$. Here, experience may play a principal role, as the designer should determine what are the variables the considered "rest" will be dependent of. It is not recommendable that a function is approximated by a constant value $\xi^{(\cdot)}$, as it is later integrated in the term $\int_0^t \Psi(t')dt'$ of the expression (4.7). This significantly deteriorates the degree of system's stability.

- If a situation arised that is not possible to identify any $\xi_j^{(k)}$ which would satisfy the preset conditions (4.6) and (4.7), then one should do the following: (i) "Narrow" the region of the state space $\underline{X}^t$ (Fig. 4.2) for which stability conditions are examined (i.e. to relax the preset constraints); (ii) Change the structure of the estimation function. For example, for the function $f(\Delta x) = \Delta x^3$ instead of the linear form $f(\Delta x) < \xi^{(1)}\Delta x + \xi^{(2)}$ take some other, more faithful approximation, say the quadratic form $f(\Delta x) < \xi^{(6)}(\Delta x - a)^2 + \xi^{(7)}$ (see Fig. 4.10); (iii) Change the gains K_P, K_V, $K_F^{(2)}$ in the feedback loop.

- If we succeeded in determining the conditions that satisfy the preset conditions of the function estimate, but if thereby the system stability conditions are not satisfied, then we have to try some of the above three possibilities to make corrections. It should be noted here once more that the stability conditions are conservative, which means that the lack of their fulfillment does not mean that the system is unstable.

- If one of the stability conditions (4.6) or (4.7) is not satisfied whereby all the coefficients $\xi_j^{(k)}$ were determined so that they satisfy the expressions in the inequalities, it is necessary to try to reduce (narrow) the region of allowed deviations $\underline{X}^t$

- The positive definite matrix H is determined from the inequality $\mathrm{H}A_1 + A_1^T\mathrm{H} \leq -2\min_{i=1,2n}|\sigma_i(A_1)| \cdot \mathrm{H}$. First, the inequality is solved for H. Most often, it happens that the solution is not a positive definite matrix. Then, it is necessary to change only the gains in the matrix A_1, and in no case the elements of the matrices H and J functioning in A_1. It is recommendable that control gains are varied so that the eigenvalues of the matrix A_1 be conjugate complex numbers and not real numbers. It appeared that only in that case the considered matrix inequality has a satisfactory solution.

The above recommendations should be borne in mind in the realization of the algorithm for determining the estimation functions as well as in the computer checking of the defined conditions of practical system's stability (4.6) and (4.7).

4.7   Conclusion

The investigations described in this chapter represent an original contribution to the analysis of vehicle dynamic performance and system stability, as well as to the issue of increasing robustness of the synthesized control laws. The test conditions for practical stability of a road vehicle interacting with its dynamic environment were derived and tests were performed for the purely positional and combined position-force dynamic control law. The test have explicitly shown the effects of various uncertainties upon the system's practical stability, allowing thus effective comparison of different control laws and study of the effects by which these uncertainties may influence a control law. The presented tests represent the first example of taking into account all the relevant aspects that can influence stability of the road vehicle in its motion. The derived stability tests represent a generalization of asymptotic (exponential) stability conditions for the dynamic control laws derived in [7] for the case of practical stability and parametric uncertainties. This means that the test derived here enables one to determine what are the conditions to be satisfied by the "environment[8]", requested nominal trajectories (w.r.t. position and force), and the control law in order to ensure practical stability of the system. Practical stability appeared to be more appropriate for this type of control tasks than asymptotic stability since the former enables one to consider the effects of uncertainties in the control laws which cannot guarantee stability at all, but can fulfil practical stability conditions (e.g. the position-force control law which cannot asymptotically stabilize the road vehicle when the environment parameters are not exactly known). Having in mind the above discussion, the main contributions of this chapter may be summarized as follows:

- A procedure for testing practical stability of a road vehicle interacting with its dynamic environment has been described for the first time in the literature.

- The practical stability test enables one to study the issue of uncertainties in the control of road vehicles without any approximation, i.e. to accurately examine the influences of model uncertainties upon the effectiveness of the different control laws and system performance.

- Since the requirements of practical stability are more relaxed and more convenient from a practical point of view than asymptotic (exponential) stability conditions, it is of essential importance to apply the practical stability test to the analysis/synthesis of the control laws taking into account the model uncertainties.

- The automatic procedure for executing the practical stability test enables scheduling of the control gains of the vehicle controller, which enhances the robustness of the control system.

The proposed stability test may be used either to check stability of a specified control law, or to establish the procedures for the synthesis of parameters of the different control laws. By this, the control law will become much more accurate and effective, i.e. the robustness of the control with respect to the uncertainties in the vehicle and environment models can be enhanced. The control robustness is one of the most relevant aspects in the tasks of road vehicle maneuvering, sudden changes of the motion direction, variation of the longitudinal and lateral accelerations, etc.

Here, only the uncertainty of parameters in the models of the vehicle and "environment" have been considered, but the approach proposed for practical stability analysis may be used to consider also the uncertainties in the models of a more general form (e.g. structural uncertainties may be analyzed, too). The presented test may seem too conservative because of the number of linearizations (approximations) that have been made. More refined approximations, e.g. by taking into account a possible dependence of the model elements (matrix of

---

[8]Note that under this notion we assume the vehicle suspensions and tire pneumatics.

inertia, Coriolis forces, Jacobian matrix, etc.) on the considered parameter d, may lead to a less conservative test. The established stability conditions may be too conservative because of their generality. Therefore, a number of special cases have to be considered aiming at deriving the stability conditions which can effectively be used in practice to analyze stability of the synthesized control laws, or to synthesize control parameters that will ensure a desired stability of the road vehicle. The practical stability test (for the position-force control law) may be improved by "decomposing" the system into the position- and force-controlled subsystems, but this is not the subject of consideration of this monograph.

Bibliography

[1] Ogata, K., Modern Control Engineering, *Prentice-Hall International Editions*, A Division of Simon & Schuster, Englewood Cliffs, N. J. 07632, 1990

[2] Stojić, M., Continual Systems of Automatic Control, *Belgrade, Scientific Book*, 1980

[3] Inagaki, S., Kshiro, I., Yamamoto, M.,"Analysis of Vehicle Stability in Critical Cornering Using Phase-plane Method", *Proceedings of International Symposium on Advanced Vehicle Control for Active Safety and Ride-Comfort-AVEC'94*, pp. 79-84, Tsukuba, Japan, October 1994

[4] Stokić, D., Vukobratović, M.,"Practical Stabilization of Robotic Systems by Decentralized Control", *IFAC Automatica*, Vol. 20, No. 3, pp. 353-359, May 1984

[5] Vukobratović, M., Stokić, D., Kirćanski, N., Scientific Fundamentals of Robotics 5: Non-Adaptive and Adaptive Control of Manipulation Robots, *Springer-Verlag*, Berlin, Heidelberg, New York, Tokyo, 1985

[6] Vukobratović, M., Stojić, R., Modern Aircraft Flight Control, Lecture Notes in Control and Information Science, Vol. 109, *Springer-Verlag*, 1988

[7] Vukobratović, M., Ekalo, Y., "New Approach to Control of Robotic Manipulators Interacting with Dynamic Environment", *Robotica*, Vol. 14, No. 1, pp. 31-39, 1996

[8] Weissenberger, S., "Stability Regions of Large-Scale Systems", *Automatica*, Vol. 9, pp. 653-663, 1973

[9] Michael, N., A., "Stability, Transient Behaviour and Trajectory Bounds of Interconnected Systems", *International Journal of Control*, Vol. 11, No. 4, pp. 703-715, 1970

[10] Stokić, D., Vukobratović, M., "Practical Stabilization of Robots Interacting with Dynamic Environment", *International Journal on Robotics and Automation*, Vol. 13, Issue 4, 1998

[11] Peng, H., Tomizuka, M., "Program on Advanced Technology (PATH) Program - Lateral Control of Front-Wheel-Steering Rubber-Tire Vehicles", *Technical Report UCB-ITS-PRR-90-5*, Institute of Transportation Studies, University of California at Berkeley, UC Berkeley, 1990

[12] Rodić, A., D., Vukobratović, M., K., "Contribution to the Integrated Control Synthesis of Road Vehicles", *IEEE Transactions on Control Systems Technology*, Vol. 7, No. 1, pp. 64-79, January 1999

[13] Rodić, A., D., Vukobratović, M., K., "Design of an Integrated Active Control System of Road Vehicles", *International Journal of Computer Application in Technology*, Special Issue on Active Structures, Vol. 13, Nos. 1/2, pp. 78-92, January 2000

[14] Stokić, D., Vukobratović, M., "Contribution to Practical Stability Analysis of Robots Interacting with Dynamic Environment", *Proceedings of the 1st ECPD International Conference on Advanced Robotics and Intelligent Automation*, pp. 693-698, Athens, Greece, September, 1995

Chapter 5

Summary and Conclusions

5.1   The Monograph Subject Matter

The subject matter considered in this monograph is modelling, synthesis, control, analysis and checking of stability of automated road vehicles interacting with the dynamic environment.

Chapter 1 describes the motivation and importance of the research in the field, along with the definition of some fundamental notions to be used later in the text. The problem of vehicle control is defined as a problem of the so-called *contact task*. Its solving, described in this monograph, represents a new approach in the field of applied research.

In Chapter 2, various planar and spatial models of road vehicles, dealt with in the literature and used in different ways in the system simulation, are described in detail along with the synthesis of the motion controllers. Among the surveyed models a central place occupies the spatial nonlinear model of the road vehicle with 22 DOFs. This model has been predominantly used in the monograph in the synthesis of the dynamic controller of the system, in the investigation of global stability, and in the simulation experiments. This chapter considers also some particular models of active systems such as active suspension system, active steering system, active driving/braking system, etc.

A new concept of the structure of an automated vehicle controller is proposed in Chapter 3. The chapter describes in detail the control block-scheme of a new, hybrid neuro-dynamic vehicle controller, designed on the basis of a nonlinear, spatial dynamic model of the system. To this purpose, use was made of a concept of centralized dynamic control based on the strategy of the so-called distributed hierarchical control on tactical and executive (operative) levels. The introduction of an additional neuro-compensator yielded enhanced robustness of the automated vehicle control system with respect to structural uncertainties of the model and to the influence of stochastic disturbances on the system. By introducing the neuro-compensator into the control scheme, a new control structure, the so-called hybrid neuro-dynamic controller, was obtained.

Chapter 4 deals with a procedure for automatic checking of the conditions of practical stability of the system vehicle-environment under the conditions of existence of variable parameters. We define the concrete expressions for stability conditions of the system considered, which enable the analysis of vehicle stability under different motion conditions. The allowed scheduling of control gains yields the enhanced adaptability of the vehicle control system to the variable conditions of motion. Also, the use of the developed procedure for checking stability and determination of control gains yields a greater robustness of the synthesized controller against the parameter uncertainties of the model used in the control synthesis.

The main contributions of the monograph will be reviewed once more in the next section.

5.2 Main Contributions

The main contributions of the monograph have been described in Chapters 3 and 4. In the frame of these chapters we proposed a novel concept of automated control of road vehicles. It differs from the solutions that can be found in the contemporary literature by its *centralized approach* to control. The majority of the works in the field of automated vehicle control are mainly concerned with the synthesis of a decentralized, partial control of the vehicle motion in particular directions of interest. Thus, independent controllers have been developed for the longitudinal and lateral vehicle dynamics and the controllers of the vehicle dynamics in the plane perpendicular to the road surface (i.e. in the yaw, roll, and pitch directions). Some of these solutions have already been in commercial exploatation in the automotive industry. However, none of them can guarantee the overall system stability simultaneosly in the six principal directions of motion. What does it practically mean? It means that in certain circumstances of motion (under the action of disturbances or in the presence of mutual coupling) the considered system can exhibit an unstable behaviour in some particular directions irrespective of the individually considered partial controllability of the object in these directions. In order to overcome such undesired situations we proposed that the synthesis should take into account the so-called centralized, overall vehicle dynamics, to which corresponds a spatial, nonlinear system model that takes care of the dynamic interactions in particular directions of motion. The application of such a dynamic model in control synthesis, whereby the conditions of system's practical stability are satisfied, ensures general stability of the system simultaneously in all six principal directions of motion. By checking the conditions of practical stability on the basis of the developed complex model of the vehicle, different control gains were synthesized that take into account certain ranges of variation of the system parameters. Therefore, it has been shown what are the conditions that can guarantee stable behaviour of the road vehicle under the assumption that the model employed describes the real system in a sufficiently accurate way. In a contrary case, the developed dynamic controller with schedulling gains should be supplied with an additional neuro-compensator whose role is to compensate for the effects of the unmodeled system dynamics such as time delays in the actuators, presence of elastic modes of the structure, Coulombs's friction at the mechanism joints, etc.

The centralized approach as a control concept adopted in the present work was realized on the basis of applying the strategy of *distributive hierachical control* on two levels - *tactical* and *executive (operative)*. In this sense, the task of the tactical control level is to take care of the overall vehicle dynamics in the course of motion, whereas the executive level, on the basis of information obtained from the tactical level, generates the control signals directly in the vehicle actuators. Thus the potential dynamic effects in the system can be described by an analytic expression representing the corresponding object model. Taking the previously determined values of control gains, obtained on the basis of checking up the conditions of practical stability, it is possible to guarantee the overall stability of the system's behaviour, even in the presence of parametric uncertainties.

As for the application of the neuro-compensator as an additional control block in the structure of the system of automatic control, its role in the proposed control scheme is to compensate for the structural inaccuracies of the system's model as well as for the unpredictable parametric uncertainties. With the aim of achieving a better control quality it has been proposed to introduce the neuro-compensator on the higher (tactical) level of the control scheme, located in the feedback loop of the control system.

Hence, the main contributions of this research monograph are related to the results that have been obtained in the synthesis of the centralized dynamic controller of the automated road vehicle; the development of computer-oriented automatic procedure for checking practical vehicle stability under the conditions of predetermined changes of model parameters and the conditions of action of the system disturbances; synthesis of the hybrid vehicle controller with the additional neuro-compensator to compensate for the effects of inaccuracies of the unmodelled vehicle dynamics. The proposed hybrid neuro-dynamic controller represents a novel conceptual solution to control, which is essentially different from the known contemporary solutions.

5.3 Applied Methodology

Control of road vehicles has been considered in the context of the application of *active systems*[1]. Also, control of an automated vehicle as a dynamic system being in the interaction with the road surface on which it moves, is considered as solving the so-called *contact task*. In this sense, the vehicle structure is treated as a mechanical system of rigid bodies which relies upon the so-called *dynamic environment*. In contrast to conventional understanding of the notion of vehicle environment assuming the physical characteristics of the road surface structure, the notion of the so-called dynamic environment takes into account the dynamics

---

[1] Control with external energy supply.

of both the suspension system (passive or active) and tire pneumatics. Therefore, when we talk about the dynamic environment of the vehicle we assume it can be either passive or active, depending on wheather it refers to a vehicle of conventional type or to the vehicle with active suspension system. If there is no any energy supply that would emit additional driving energy to the "environment" with the aim of chaging the values of its parameters, then we speak of the so-called *passive dynamic environment.* In the majority of examples, a passive dynamic environment exhibits unchangeable dynamic behaviour because the values of its dynamic environment parameters are constant or slightly changeable quantities. If there are technical possibilities to change the values of dynamic environment parameters, then we speak of the so-called *active dynamic environment.* In order to have such an active environment it is necessary to feed energy from an external source in an amount sufficient to change the values of the environment parameters. In this way it is possible to change the dynamic behaviour of the environment itself in accordance with the concrete needs. In Fig. 5.1 is presented a relatively simple two-dimensional model of non-dynamic and dynamic environments in the interaction with a dynamic system.

Taking into account what was said above, the conventional spatial model of the vehicle (Fig. 2.13) has been apparently partitioned into two dynamic subsystems: *vehicle body (structure)* and *active dynamic environment.* By adopting such a concept of the system considered and possibility of its decoupling into the vehicle structure and dynamic environment we were able to synthesize different dynamic algorithms of control on the tactical level. These are the algorithms based on the purely positional (3.3) and combined position-force control (3.8). In doing so, the known *inverse dynamic method* was employed as the position control law. As far as the estimation of variable system parameters is concerned we applied the *recursive least squares deviations method* of estimation, which is often used in the tasks of real-time parameters estimation.

The methodology used in the synthesis of the neuro-compensator is concerned with the application of an artificial multilayered neural network, which was trained using the *standard back propagation method.* This method gave satisfactory simulation results in the implementation in the frame of the proposed structure of the *hybrid neuro-dynamic vehicle controller.*

5.4   Further Research Directions

The theoretical results presented in this monograph should serve as a basis for practical realization of a system for automatic control of motor vehicles. Namely, we think that the conventional power-driving systems (internal-combustion engine) should be replaced with the so-called *hybrid four-wheel drivetrain,* as this yields a better controllability of the system and a lower energy consumption. In the development of the proposed concept of automatic vehicle control presented in this monograph, more attention should be paid to the future synthesis of control on the operative level. This would allow taking into account all the concrete construction characteristics of the road vehicle on the level of the particular con-

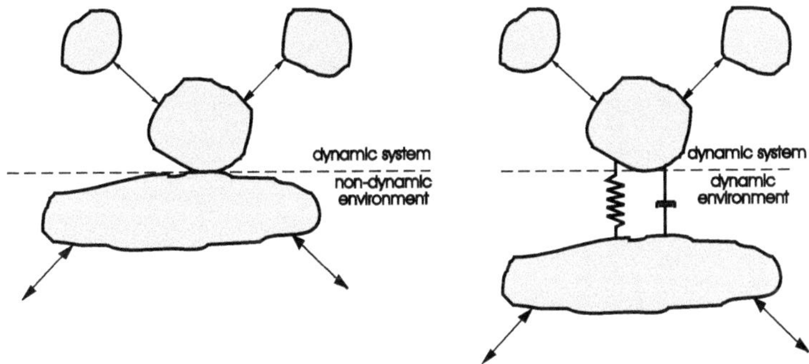

Figure 5.1: Model of non-dynamic and dynamic environment in the interaction with a dynamic system

trol subsystems. In this way the vehicle control performance would be advanced as the control synthesis would also take into account the corresponding design characteristics. It is known that these characteristics may vary from model to model in dependence of the concrete dedication of the vehicle. Hence, different types of active systems, whose constant technological improvements lead to enhanced technical capabilities, could be included into the existing control structure of the vehicle controllers. It is essential to point out that the proposed control scheme should not be changed on the tactical control level but it can be upgraded and bettered only on the executive level. Of course, this gives full freedom to the manufacturers in the automotive industry to propose diverse solutions for the vehicle actuators.

Apart from the above directions of the possible further development, we think that testing of some new intelligent control techniques could also potentially advance the overall quality of vehicle motion handling on tactical level. In dependence of the newly proposed algorithmic solutions, appropriate hardware solutions of the controller structure should be developed with the aim of ensuring the necessary microprocessor support of the vehicle control.